SpringerBriefs in Philc

For further volumes:
http://www.springer.com/series/10082

Christian Beenfeldt

The Philosophical Background and Scientific Legacy of E. B. Titchener's Psychology

Understanding Introspectionism

Christian Beenfeldt
Department of Media, Cognition and
Communication
University of Copenhagen
Copenhagen
Denmark

ISSN 2211-4548 ISSN 2211-4556 (electronic)
ISBN 978-3-319-00241-5 ISBN 978-3-319-00242-2 (eBook)
DOI 10.1007/978-3-319-00242-2
Springer Cham Heidelberg New York Dordrecht London

Library of Congress Control Number: 2013935489

Printed on acid-free paper

Springer is part of Springer Science+Business Media (www.springer.com)

Acknowledgments

For their invaluable comments on the research and on the arguments in this book—throughout their various stages of evolution—I must first and foremost thank Daniel N. Robinson and Bill Child. I also thank Martin Davies, Tim Bayne, Tim Crane, Finn Collin, Dan Zahavi, and Darryl Wright for their keen observations and critical comments.

Individual components of this material have been presented at a number of different venues. These include the British Society for the Philosophy of Science, the Joint Session, the Pittsburgh HPS department, UC Irvine, the University of Copenhagen, the Danish Society for Philosophy and Psychology, and, most recently, the ASSC 16. I would like to thank the participants at these events for stimulating feedback on my work.

The writing of this book has been made possible by a generous Carlsberg Foundation Research Grant.

Contents

Part III The Preeminence of Analysis, Not Introspection

Introduction

Psychology, like every other science and, indeed, like every other field of human enquiry, did not spring fully grown and fully armed from Zeus' forehead. As an experimental science, the field's past stretches back beyond the so-called cognitive revolution of the 1950s and 1960s and, earlier still, beyond the so-called behavioral revolution of the 1910s and 1920s.[1] Beyond this, of course, reflections upon the nature of the human psyche predates antiquity. Recognizing that the lessons of psychology's past provide perspective, reflexivity (Smith 2010), and perhaps even the impetus for a reevaluation of some aspects of current practice, we evidently need to *get those lessons right*.

With this in mind, consider the well-trodden story about the coming-to-be of modern psychology. Repeated in introductory psychology textbooks, this narrative is a core disciplinary legend. It usually goes something like this:

> (1) Psychology, as instituted in the universities, began as the study of mind, *based, almost exclusively, on the method of introspection*.
> (2) In reaction to the *blatant unreliability of the introspective method*, behaviourism then redefined psychology as the study of behaviour, based, primarily, on the objective method of experimentation.
> (3) In reaction to the limited research agenda and theoretical bankruptcy of behaviourism, the 'cognitive revolution,' in turn, restored the mind as the proper subject of psychology (but now with the benefit of the rigorous experimental and statistical methods developed within behaviourism) (Costall 2006, p. 635).[2]

The above quote is Alan Costall's summary of the standard story. The story, as Costall sees it, is mere mythology—and it forms part of the widespread "fictional history" of psychology (Costall 2006, p. 635). Mythical or not, the story has been a long-standing fixture of psychology's scientific identity. Already in the early 1990s, for example, Thomas Leahey published a trenchant criticism of this myth in *American Psychologist*. Leahey's target was a

[1] See Mandler 2002 for a nuanced discussion of the extent and influence of behaviorism as well. Mandler also argues for the gradual, not revolutionary, transition from behaviorism to cognitive psychology.

[2] Italics added to this quote. Parenthesis in original. British spelling retained.

> story of the development of American psychology widely told and widely repeated. In the beginning—1879—psychology was born as *the science of mental life, studying consciousness with introspection.* Then, in 1913, the dominance of mentalism was challenged and shattered by the rude and simplistic behaviorists … However, in 1956, a new revolution began, its makers waving the banner of cognition, aided by outside forces from linguistics and artificial intelligence (Leahey 1992, p. 308, italics added).

The aim of this book is to reopen and to rewrite, in part at least, the first chapter of this history of psychology. Contrary to the contention that prebehavioristic introspective psychology—or, *introspectionism*, as we will refer to it here—was undone by its overreliance on introspection in the study of mental life, it will here be argued that the major philosophical flaw of introspectionism was its utter reliance on key *theoretical* assumptions inherited from the intellectual tradition of British associationism, assumptions that were upheld in *defiance* of introspection.

This thesis sounds paradoxical to be sure. How could a historical movement known as *introspection*ism not have been profoundly committed to the use of introspection? That's a good question—and, in fact, this entire book can be seen as an endeavor to address precisely this query in a comprehensive and scholarly manner.

For now, the following introductory clarification must suffice. The term "introspectionism," although widely used by philosophers, psychologists, and historians for almost a century, is not as unproblematic as it might appear. Typically, the term is used to refer to a family of influential nineteenth and early twentieth century systems of experimental psychology, most notably the psychology of Edward Bradford Titchener. Yet, it is important to realize that the term was attached to these systems in general, and to Titchener's system in particular—*by their critics*.

As Kurt Danziger has argued, "no proponent of introspection as a basic method of psychology ever called himself an introspectionist" (Danziger 1980, p. 241). Indeed, the "very notion of an 'introspectionist psychology' is a product of behaviorism" (Danziger 1980, p. 241). The great historian of psychology, Edwin G. Boring, made largely the same point almost 20 years earlier, observing that

> [i]ntrospection got its ism because [the] protesting new schools needed a clear and stable background against which to exhibit their novel features. No proponent of introspection as the basic method of psychology ever called himself an introspectionist. Usually he called himself a psychologist (Boring 1963, p. 172).

Titchener, along with a number of other thinkers at the time, were thus labeled and, to borrow a term from political life, "negatively campaigned," as inveterate practitioners of old-fashioned "introspectionism"—making the parvenu school of *behavior*ism look more appealing by comparison. The name "introspectionism" has stuck. This, in itself, is not too troubling. After all, "what's in a name?"—as Juliet put it so poetically about sweet Romeo. The real problem lies in the supposition, widely held today, that introspectionism is the preeminent example in the history of science of a psychological system built on a fundamental commitment to introspection. This estimation, as we shall see, is confused and inaccurate.

Danziger has noted that, for "a topic of rather central importance in the emergence of modern psychology, introspection has not been accorded the *historical* attention it deserves" (Danziger 1980, p. 241, italics added). Perilous consequences follow from this inattention. This book traces the intellectual background of introspectionism in early modern philosophical thought (Part I). On this basis, we will proceed to examine the nature of Titchener's distinctive system of psychology and to argue that it is really a form of associationism (Part II). Building on this two-stage analysis, the argument will be made that, contrary to widespread and popular opinion, introspectionism had very little to do with what we today would recognize as introspection (Part III).

In sum, the aim of this book is to offer a fundamental reconceptualization of Titchenerian introspectionism with respect to its basic philosophical and scientific stance, its investigative methodology, and the actual cause of its demise.

References

Costall A (2006) 'Introspectionism' and the mythical origins of scientific psychology. Conscious Cogn 15(4):634–654

Danziger K (1980) The history of introspection reconsidered. J Hist Behav Sci 16(3):241–262

Leahey TH (1992) The mythical revolutions of American psychology. Am Psychol 47(2):308–318

Mandler G (2002) Origins of the cognitive (r)evolution. J Hist Behav Sci 38(4): 339–353

Smith R (2010) Why history matters. Hist Philos Psychol 12(1):26–38

Part I
Intellectual Background

Chapter 1
Early British Associationism

1.1 The Psychology of Empiricism

We begin our analysis by considering the philosophical school of empiricism in early modern thought.[1] In a sentence, a central feature of this philosophical approach was the epistemological contention that all knowledge has its origin in simple sensory experience. This claim raises the following *psychological* question: given that complex mental phenomena (such as an abstract train of thought) obviously are not direct and simple deliverances of sense, what is their nature and where do they come from? Associationism was the name given to the increasingly elaborate account developed to answer this query.

"British empiricists are often called the British associationists," as George Mandler has put it, "because their work is based on a fundamental principle of mental life—the association of ideas" (Mandler 2007, p. 18). Titchener similarly observed that the

> association of ideas itself came to be the guiding principle of the British school of empirical psychology. So well did it work, as an instrument of psychological analysis and interpretation, that Hume compared it to the law of gravitation in physics … [a]ll the great names in British psychology, from Hobbes down to Bain, are connected with this doctrine of the association of ideas (Titchener 1926, p. 374).

Let us begin our study of associationism with Thomas Hobbes (1588–1679), who was, at the dawn of the modern age, the fertile parent of so many influential intellectual offspring, and who has variously been called the father of empiricism, the father of empirical psychology, and the father of associationism—all with good reason (Klein 1970, p. 323; Pillsbury 1929, p. 71). For most of his long life, he was tutor to the Cavendish family, and he only began to produce substantial

[1] According to Danziger, it is precisely the "rise of an empiricist philosophy at the end of the seventeenth century" to which the "origins of the modern concept of introspection [was] closely tied" (Danziger 2001, p. 7889).

C. Beenfeldt, *The Philosophical Background and Scientific Legacy of E. B. Titchener's Psychology*, SpringerBriefs in Philosophy, DOI: 10.1007/978-3-319-00242-2_1,

original work in philosophy relatively late in life. Major influences on Hobbes' thinking included Euclid and the deductive methodology of geometric proof in his *Elements*, as well as Galileo's contemporary work in physics.

1.2 Thomas Hobbes

Hobbes offered a distinctive philosophical theory according to which all of reality (including human beings and their mental life) is to be accounted for in terms of matter-in-motion and the deductively demonstrated ramifications of this fact.[2] As a distinguished historian of psychology has put it,

> [w]ith Descartes, Hobbes identified matter with extension. Siding with the Gassendists, he insisted that only body can affect body and that only *matter in motion* can serve as the subject of scientific inquiry (Robinson 1986, p. 305).

Hobbes's masterpiece was the 1651 work *Leviathan*,[3] where the state is presented as a titanic organism consisting of individuals who, in their desire to achieve peace and security, all contractually vest their right to the use of force in a single mighty sovereign. Serving as the philosophical groundwork for this vision of government,[4] we find in the first nine chapters of *Leviathan* a seminal account of knowledge and of the mind.

In the world of *Leviathan* the "thoughts of man" are either (1) sense data or (2) content "derived from that original" (Hobbes 1994, p. 6). The former are caused by an external body imparting motion to the appropriate sensory organ, setting off a complex chain-reaction of internal pressure and counter-pressures in our nerves, membranes, brain and heart; light, sound, odor, flavor, heat, cold, hardness, and softness are thus all simply to be thought of as motions of matter imparted to our sensory receptors (Hobbes 1994, pp. 6–7).[5] Derivative content includes imagination, memory and dreams, which are "decaying sense," understood as a continuing internal motion produced by the motion of original sensory stimulation (Hobbes 1994, p. 8). Understanding is the state of imagination produced by the stimuli of "voluntary signs" such as words (Hobbes 1994, p. 11), reasoning is the act of performing arithmetic ("reckoning") with words (Hobbes 1994, pp. 22–3), and truth is a matter of fitting those words into the right sequence (Hobbes 1994, p. 19).

---

[2] In Robinson's words, "[i]t was Hobbes' belief that a science of society could be established with the same rigor and sureness enjoyed by Mechanics" (Robinson 1986, p. 303).

[3] In full, *Leviathan Or the Matter, Forme and Power of A Common Wealth Ecclesiasticall and Civil.*

[4] It is an open question whether Hobbes' mechanism was substantially shaped by his political philosophy or not. In an analysis that draws broadly upon Hobbes' work, Gray (1978) argues that Hobbes should be seen as a truly systematic mechanist.

[5] In reality, in Hobbes' view, those qualities are "seeming and apparitions only;" "the things that really are in the world without us, are those motions by which these seemings are caused. And this is the great deception of sense ..." (Hobbes 2008, p. 26).

Unlike the lower animals, human beings are capable of undergoing a law-like succession of one thought to another that may be termed a *train of thoughts* or imaginations (Hobbes 1994, p. 12, italics added), and although we may not be certain of what comes next in a train of thought, we *can* be certain that "it shall be something that succeeded the same before, at one time or another" (Hobbes 1994, p. 12) So, for example, in

> a discourse of our present civil war, what could seem more impertinent than to ask ... what was the value of a Roman penny? Yet the coherence to me was manifest enough. For the thought of the war introduced the thought of the delivering up the king to his enemies; the thought of that brought in the thought of the delivering up of Christ; and that again the thought of the 30 pence which was the price of that treason; and thence easily followed that malicious question; and all this in a moment of time, for thought is quick (Hobbes 1994, p. 12).

We can further differentiate unregulated trains of thought (such as a daydream) from trains of thought that are regulated by some desire (e.g. cause-to-effect reasoning), and this provides us with the basis for accounting for human language, inasmuch as the "general use of speech" is to "transfer the train of our thoughts into a train of words" (Hobbes 1994, p. 16). The Hobbesian notion of human cognition as a law-like associative process working on what ultimately are sensory elements prefigures, as we shall see, the essence of associationism.

1.3 John Locke

The next high point in early associationism was the *Essay Concerning Human Understanding* completed by John Locke (1632–1704) in 1690. Locke was not only the central figure in British empiricism; he was also a pioneer of association psychology. Like Hobbes, he regarded the work in physical science of his time as a model of great achievement.[6]

According to Locke, our minds begin as tabula rasa and our subsequent knowledge all derives from experience.[7] Experience, in turn, comes to us from two basic sources: sensation and reflection (Locke 1979, pp. 104–5).

> Let us then suppose the Mind to be, as we say, white Paper, void of all Characters, without any Ideas; How comes it to be furnished? ... To this I answer, in one word, From Experience: In that, all our Knowledge is founded; and from that it ultimately derives it self. Our Observation employ'd either about external, sensible Objects; or about the

[6] A model with which he was personally familiar. During his years at Christ Church College in Oxford, for example, Locke became well acquainted with many of the great scientists of the seventeenth century—including Robert Boyle, John Wilkins, Thomas Willis, John Wallis, Robert Hooke, David Thomas, and Richard Lower. He collaborated in experimental work with several of these great figures (Rogers 1978, p. 223).

[7] Strictly speaking, as Yolton reminds us, "no idea *comes from experience* on Locke's program since it is ideas of all sorts which *make up* or *constitute* experience" (Yolton 1963, p. 53).

internal Operations of our Minds, perceived and reflected on by our selves, is that which supplies our Understanding with all the materials of thinking. These two are the Fountains of Knowledge, from whence all the Ideas we have, or can naturally have, do spring (Locke 1979, p. 104 italics removed).[8]

Sensation provides us with the ideas we have of yellow, soft, bitter, and so on.[9] Reflection—or, introspection as we, following Gilbert Ryle, might term it[10]—provides us with "perception of the operations of our own mind" and delivers such ideas as perception, thinking, doubting, believing, reasoning, knowing, willing and so on. "The Understanding seems to me," Locke wrote, "not to have the least glimmering of any Ideas, which it doth not receive from one of these two" (Locke 1979, p. 106). These are then the basic building blocks of mental content,[11] and complex ideas are the compounds produced therefrom.[12]

These simple Ideas, the Material of all our Knowledge, are suggested and furnished to the Mind, only by … Sensation and Reflection. When the Understanding is once stored with these simple Ideas, it has the Power to repeat, compare, and unite them even to an almost infinite variety, and so can make at Pleasure new complex Ideas (Locke 1979, p. 119, italics removed).

In short, we have here a reductionist account of mind, in which mental content is reducible to various ideas of sensation or reflection.[13] The ideas of sensation, in

---

[8] See Klein (1970, pp. 360–368) for a discussion of why Locke was opposed to innate ideas but not also to congenital experiential content derived from the exercise of the senses in the womb.

[9] On a robustly Newtonian account, these are sorts of "atoms" (Smith 1987, p. 125).

[10] The relevant passage from Ryle reads as follows: "Locke … described the observational scrutiny which a mind can from time to time turn upon its current states and processes. He called this supposed inner perception 'reflection' (our 'introspection'), borrowing the word 'reflection' from the familiar optical phenomenon of the reflections of faces in mirrors. The mind can 'see' or 'look at' its own operations in the 'light' given off by themselves. The myth of consciousness is a piece of para-optics" (Ryle 2000, p. 153).

[11] In Book IV Locke makes certain, arguably more Cartesian, statements that seem at odds with the earlier body of work in the *Essay*. Klein argues that "[m]any chapters separate Book I from Book IV, and considering that 20 years elapsed before the *Essay* was completed, the books were probably also separated by many years. As a consequence, by the time Locke set to work on Book IV he may no longer have been vividly aware of the details of the arguments presented in Book I. At all events, it is difficult to reconcile what he [says about intuitive knowledge of one's own existence] with his earlier repudiation of innate ideas" (Klein 1970, p. 373). No position is taken on this issue in the present work. We are here concerned with the familiar Lockean view from the first books of the *Essay*.

[12] For a discussion of how Locke's view of the mind seems to have been informed by atomism, see Anderson (1965). See Soles (1985) for a discussion of how Locke's view of simple ideas fit, via the notion of "body" (and Locke's familiarity with the microscope), with the atomic view of matter.

[13] Locke treats "ideas" as something of a mental catch-all. "Whatsoever the Mind perceives in it self, or is the immediate object of Perception, Thought, or Understanding, that I call Idea (Locke 1979, p. 134). In one examination of Locke's uses of "idea" in the *Essay*, the term was found to apply to what today might be called percepts, concepts, belief, knowledge, and also qualia (Nathanson 1973, pp. 29–36). "Idea" is here conceived of as an object of thought, as opposed to an act or a disposition (McRae 1965, p. 175).

turn, are understood as simple, atomic units that can be compounded into increasingly more complex ideas[14]—and if we enquire into the nature of these elements we find that these are discrete items such as "Yellow, White, Heat, Cold, Soft, Hard, Bitter, Sweet, and all those which we call sensible qualities" (Locke 1979, p. 105, italics removed). Unlike Hobbes, Locke recognizes reflection as a separate and basic source of mental content, but the disagreement is much less substantive than it first appears. In Locke's theory, the introspective source of mental content is accounted for in much the same way that the sensory source of content itself is accounted for. Indeed, Locke famously speaks of the former source as an *internal sense.*[15]

> The Other Fountain,[16] from which Experience furnisheth the Understanding with Ideas, is the *Perception of the Operations of our own Minds* within us, as it is employ'd about the ideas it has got; which Operations, when the Soul comes to reflect on, and consider, do furnish the Understanding with another set of Ideas, which could not be had from things without: and such are, Perception, Thinking, Doubting, Believing, Reasoning, Knowing, Willing, and all the different actings of our own Minds; which we being conscious of, and observing in our selves, do from these receive into our Understandings, as distinct ideas, as we do from Bodies affecting our Senses. This Source of Ideas, every Man has wholly in himself: And though it be not Sense, as having nothing to do with external objects; yet it is very like it, and might properly enough be call'd internal sense (Locke 1979, p. 105, italics removed).

We turn now to the mechanism of association itself. The famous chapter on this topic in the *Essay* "On the Association of Ideas"[17] is often credited with providing both inspiration to and, subsequently, a name for associationism, but it is less often recognized that Locke was here not in the least interested in the general psychological laws of operation that are characteristic of healthy mental functioning. Rather, he treats the mental mechanics of association as way to account for what we would today term irrationality and psychopathology. Among the cases he discusses is that of a man "perfectly cured of Madness by a very harsh and offensive Operation" who could no longer stand the sight of the surgeon because it brought back the "Idea of that Agony which he suffer'd from his Hands" (Locke 1979, p. 399, italics removed). He also considers the anecdotal case of a young man who learned to dance while there was an old trunk in the room, and subsequently could only keep up the performance with the said accoutrement nearby.[18] Regarding

---

[14] Concerning the simple ideas from sensation or reflection, Anderson explains that "both kinds of simple ideas are, as it were, *given* to the mind. That is, the mind can neither make new simple ideas nor can it destroy those it has" (Anderson 1965, p. 206).

[15] Gertler (2011, Chap. 2) has a helpful discussion of Locke's view of introspection and the inner sense.

[16] Other than sensation.

[17] Book II, chapter XXXIII. This chapter was added to the fourth edition of the *Essay* in 1700 (Ferg 1981, p. 173).

[18] Apparently, the "Idea of this remarkable piece of Household-stuff, had so mixed it self with the turns and steps of all his Dances that though in that Chamber he could Dance excellently well, yet it was only whilst that Trunk was there ..." (Locke 1979, p. 399, italics removed).

irrationality in general, Locke counsels against perilous associations of ideas, "wrong and unnatural Combinations," that could establish "Irreconcilable opposition between different Sects of Philosophy and Religion" (Locke 1979, p. 400). In his estimation, ideas that are wrongly connected and fused together, are "the foundation of the greatest, I had almost said, of all the Errors in the World" (Locke 1979, p. 400).

Despite Locke's apparent lack of interest in providing a systematic account of the healthy human mind, David Hartley observed that the "influence of association over our ideas, opinions, and affections, is so great and obvious, as scarcely to have escaped the notice of any writer who has treated of them, though the word *association*, in the particular sense here affixed to it, was first brought into use by Mr. Locke" (Hartley 1834, p. 41).

1.4 David Hume

In focusing on David Hume (1711–1776), the last of the three early associationist thinkers we will discuss, we move from the view that the psychological mechanism of association is at the root of irrationality and madness, to the view that it is the very source of our experience of ontological order rather than chaos. Hume is often seen as the author of what amounts to a profoundly skeptical assault on knowledge, certainty, induction and even causality, but what we must here emphasize is that he fused such skepticism with the positive reinterpretation of these notions as expressions of *psychological law*. To Hume, psychological law meant the basic laws or regularities of association (Bricke 1974).

We begin with the familiar British empiricist assumptions. Mental content is divided into two basic types, impressions and ideas, according to differences in their degree of experienced "force and liveliness" (Hume 1978, p. 1).[19] Impressions, characterized by their rich force and liveliness, encompass all of our sensations, emotions and passions. Ideas are the "faint images" of the impressions—a "copy taken by the mind"—that we may encounter in cognitive processes such as thinking, although they are always the derivatives of the "correspondent" original impressions (Hume 1978, pp. 1–8). The relationship is a one-to-one mapping inasmuch as "every simple idea has a simple impression, which resembles it; and every simple impression a corresponding idea" (Hume 1978, p. 3).

Mental units come in simple or complex form. In simple form, they "admit of no distinction or separation" whereas in complex form they can be separated into constituent parts (Hume 1978, p. 2). As with Hobbes and Locke, we again have an essentially associationistic account and we find sophisticated mental content

[19] Setting aside, here, some cases in which they may be difficult to distinguish, such in sleep, fever, madness, bouts of violent emotions, or instances in which "our impressions are so faint and low, that we cannot distinguish them from our ideas" (Hume 1978, p. 2).

(ideas) reductively accounted for as a derivative type ("faint images") of the sensory simples (impressions). Following Locke, Hume also divides impressions into sensation and reflection, although the difference between the two is slight: if the proximate source for an impression is a "copy" (i.e. an idea) of another impression (such as the idea of pain producing the impression of aversion) we may call this reflection, if there is no such proximate source, we may call it a sensation.[20] Ultimately, however, we always find an initial impression as the first link in the chain (Hume 1978, pp. 7–8).

Consider now the laws of association, as accounted for in Sect. 1.4 of Hume's Treatise ("Of the connexion or association of ideas"). Similar to Newton's three laws of motion, Hume suggests (on first pass) three "universal principles" (Hume 1978, p. 10) that are responsible for a type of "attraction, which in the mental world will be found to have as extraordinary effects as in the natural" (Hume 1978, pp. 12–13). These principles explain the non-random generation, i.e. the aggregation, of complex ideas but, unlike gravity, they disclose only a "gentle force, which commonly prevails" (Hume 1978, p. 10). The three universal principles are resemblance, contiguity in space and time, and causality.[21] Following this line of thought, we are led to a view of the mind as

> a kind of theatre, where several perceptions successively make their appearance; pass, re-pass, glide away, and mingle in an infinite variety of postures and situation (Hume 1978, p. 253).

Turning to Hume's metaphysics, we find the claim that the principle of causality is reducible to habitually experienced contiguity. However, this, of course, is significant. It implies a robust form of anti-realism in which all causal interactions in reality (e.g. flames *causing* dry paper to burn, germs *causing* fermentation or disease, the heart *causing* blood to circulate) are now recast as mere dispositional prejudices of the human psyche, gearing us to the interminable misapprehension of propinquities as necessities. What we normally think of as a basic ontological fact—the causal nature of external reality itself—is now viewed as having its genesis in the workings of human psychology.[22] Debate in Hume scholarship about the level of skepticism advanced by Hume has raged for centuries. The "Beattie-Reid tradition of interpretation" of Hume as an arch-skeptic is, perhaps, less popular

[20] As Hume explains, this kind of impression "arises in the soul originally, from unknown causes" (Hume 1978, p. 7) and the "examination of our sensations belongs more to anatomists and natural philosophers than to moral; and therefore shall not at present be enter'd upon" (Hume 1978, p. 8).

[21] Hume may have taken these categories from the work of George Berkeley (Warren 1967, p. 45). For a more detailed discussion see Bricke (1974).

[22] Hume also attacked the notions of the self and personal identity (most notably in Hume 1978, pp. 251–263). James points out how this follows from the associationism shared by Hume and other subsequent empiricists (such as James and John Stuart Mill, whom we will consider shortly). They "have thus constructed a psychology without a soul by taking discrete 'ideas,' faint or vivid, and showing how, by their cohesion, repulsions, and forms of succession, such things as reminiscences, perceptions, emotions, volitions, passions, theories, and all the other furnishings of an individual's mind may be engendered. The very self or ego of the individual comes in … as their last and most complicated fruit" (James 1952, p. 1, italics removed).

today than it was three decades ago—but, as Louis Loeb observes in his contribution to *Synthese* on Hume's argument about induction, "[n]o one denies that Hume at least takes us to the brink of deeply skeptical results" (Loeb 2006, p. 322).[23]

1.5 Elementism, Reductionism, Sensationism, and Association

The theorizing of Hobbes, Locke, and Hume forms part of what what we can call the early phase of associationism. Already at this early point, however, we can recognize[24] four distinctive and fundamental assumptions informing this school of psychological thought:

1. Elementism
2. Decompositional reductionism
3. Sensationism
4. The laws of association.

Elementism is the doctrine that a given system is composed, without significant residue, of a finite number of ultimate, indivisible, atomic constituents (usually limited to a few types) that combine to produce complex wholes. In Hume's system, the impressions are the basic elements and the ideas are to be understood as the derivatives, the faint images thereof.[25] In Locke's system the ideas of sensation are understood as elementary units that can be compounded into increasingly more complex ideas. This notion is similar, in the psychological realm, to the mechanical hypothesis in the Enlightenment philosophy of nature. Here, as Rogers notes, with the assumption of

> a limited number of basic corpuscles of various shapes, moving at variable speeds, we can account for the vast number of varying properties we discover in the world in rather the same way that the limited number of letters of the alphabet can account for all the works of literature written in various languages (Rogers 1996, p. 53).

Decompositional reductionism is the kindred doctrine according to which any (non-elemental) level of a given system can be fully accounted for by

---

[23] We should note the tension between Hume's commitment to (1) a Newton-inspired psychology and (2) the skeptical project that (especially on a Logical Positivist interpretation) calls into question the very meaningfulness of items such as invisible causal forces (e.g. gravity), imperceptible entities (e.g. physical atoms), and so on.

[24] The claim is not that these three early empiricist thinkers all equally embraced what later became the mature doctrine of associationism. Drever, for example, contends that Locke was "equivocal" on elementism while Hume "spoke as if no other view [than elementism] was possible" (Drever 1965, p. 126). The point is that there was a substantial *prefiguring* of associationism in Hobbes, Locke, and Hume. In the writings of Hartley, James Mill, and John Stuart Mill, as we shall see, the doctrine becomes fully elaborated.

[25] Boden (2006, p. 125) uses the phrase "atomistic" for Hume's notion of mental associations.

decomposition of that level into simpler complexes, still simpler complexes and, ultimately, to the irreducible building-blocks or constitutive elements of that system.[26] In Hobbes's system we find that the thoughts of man, if they are not in fact sense data, are "derived" from sense data. Imagination, memory and dreams are similarly accounted for as "decaying sense," i.e. ultimately in terms of the state or condition of sense data. In Locke's system, too, we find that all mental content ultimately can be reduced to ideas of sensation or of reflection.[27]

Elementism and decompositional reductionism are closely related and can be seen as two sides of one metaphysical coin.[28] Elementism is the claim that the system is constructed, bottom-up, from of the atomic constituent input alone. Decompositional reductionism is the claim that the complex levels of the system can be decomposed top-down into simpler complexes and ultimately to the most basic and simple constituents.[29] The "whole historic doctrine of psychological association," as William James famously observed, is tainted

[26] It is anachronistic to impose terminology developed several centuries later in modern philosophy of science on these notions. It might be pointed out, however, that the form of reductionism discussed here is similar to *methodological reductionism* in its emphasis on explaining the macroscopic by means of the microscopic—i.e. on its general methodological strategy of finding more minute entities (in this case, *mental* entities) to explain the grossly observed data. It is similar to *ontological reductionism* in its general conviction that, within the domain of the mental, everything is to be understood as consisting of the posited mental elements, their aggregated collections, and their causal interactions (leaving open whether there are ontologically *real* forces causing these interactions or whether the aggregations are ultimately to be understood as brute statistical regularities).

[27] This is not meant to suggest a strong commitment to materialist corpuscularianism on Locke's part; the interpretation of Locke's psychology advanced here is not necessarily at odds with a reading of Locke as a defender of phenomenal explanations in experimental physical science (a view advanced in Gaukroger 2009).

[28] It is logically possible to be a decompositional reductionist without also being an elementist. For example, someone could think that the decomposition, while theoretically possible, never bottoms out at terminal elements but continues as an interminable series. Also, even if one grants a terminal point to the reduction, this may be conceived not as an atomic unit but, say, as a field. Finally, one might (as we shall see John Stuart Mill in fact does) introduce the idea of mid-level fusion events, allowing one to retain elementism while only embracing a slightly attenuated form of decompositional reductionism.

[29] Because repeated use of the terms "reduction" and "reductionism" will be made in the following analysis, a disclaimer should be offered. The broad concept of reductionism has attracted much attention in philosophy of science and philosophy of biology circles in recent decades, while little agreement has been reached, on even a basic definition. As Nagel put it several decades ago, "[a]lthough the term 'reduction' has come to be widely used in philosophical discussions of science, it has no standard definition. It is therefore not surprising that the term encompasses several sorts of things which need to be distinguished" (Nagel 1970). This is still largely true. The present work takes no position on the general question of reductionism in science. It is merely concerned with the specific, decompositional form of reductionism characterized above. The terms "decompositional reductionism" and, simply, "reductionism" will henceforth be used interchangeably but always in close connection with the doctrine of elementism.

with one huge error—that of the construction of our thoughts out of the compounding of themselves together with immutable and incessantly recurring "simple ideas." It is the cohesion of these which the "principles of association" are considered to account for (James 1952, p. 362).

Sensationism[30] is the doctrine that normal human sense experience consists of sensation, understood as discrete experiential units such as sound, odor, flavor, heat, cold, hardness, and softness. The associationists differed in the units selected as candidates, but the common idea is that there is some finite number of discrete sensation-atoms that serve as the ultimate building blocks of the human mind. In early associationism, this is most clearly the case with Locke, whose elements are discrete sensory items such as yellow, white, heat, and cold.[31] In this way, sensationism establishes for associationism the identity of the bottom-up constituents of the system.[32]

Briefly put, the laws of association express the notion that human mental life is governed by certain universal laws of attraction that are responsible for the non-random connection of mental items. We find this first (if only in a somewhat preliminary form) in Hobbes's notion of the "train of thought" process, which always conforms to the lawful regularity of producing something that "succeeded the same before, at one time or another" (Hobbes 1994, p. 12). The preferred parallel to the laws of association was the notion of a universal attractive force, proportional to the mass product and inversely proportional to the square of the distance between the masses, and the laws of association, as we shall see, were routinely thought of as identifications of comparable universal regularities in the mental realm. Hume explicitly makes the comparison to the Newtonian gravitational force. For Locke, as we saw, ideas associated together wrongly can even cause events on a world historical scale, such as irreconcilable religious conflict.

[30] Sensationism is a strong philosophical claim about the nature of human (mainly sensory-perceptual) awareness. The claim is that this awareness is to be understood in terms of discrete and atomic sensory units. Psychophysics studies the relationship between stimulus magnitudes and the magnitudes of conscious effect. One widely used notion in psychophysics is that of the just noticeable difference (JND), understood as the smallest difference stimulus magnitude that a percipient can reliably detect. This is an experimentally established response relationship that, by itself, leaves open a wide range of theoretical positions on the ultimate nature of human psychology. Modern signal detection theory (SDT) has emphasized the role of decision-making factors (e.g. response bias) in psychophysical detection tasks, and those findings do tend to speak strongly against interpreting JNDs as simple sensory "atoms".

[31] One of the most familiar distinctions in Lockean epistemology is the one between the primary qualities (e.g. occupying space, being in motion or at rest, having solidity and so on) that the object has independent of us and the secondary qualities (color, taste, smell and so on) that are produced in us by the objects. Within associationistic thought, however, the primary qualities tend to migrate into the category of secondary qualities. And so, as we shall see, James Mill reduced even primary qualities such as hardness, extension, and weight to simple sensations and suggests that motion, time and space itself may also be subject to such a reduction.

[32] Logically speaking, of course, one can be an elementism without also being a sensationist; sensationism is one possible way of fixing the identity of the smallest mental constituents.

We turn now to the mature phase of the associationist movement, where the central premises were developed into a distinct philosophico-psychological school of thought.

References

Anderson RF (1965) Locke on the knowledge of material things. J Hist Philos 3(2):205–215

Boden M (2006) Mind as machine: a history of cognitive science, vol I. Clarendon Press, Oxford

Bricke J (1974) Hume's associationist psychology. J Hist Behav Sci 10(4):397–409

Danziger K (2001) Introspection: history of the concept. In: Smelser NJ, Baltes PB (eds) International encyclopedia of the social and behavioral sciences. Elsevier Science, Pergamon, pp 7888–7891

Drever J (1965) The historical background for national trends in psychology: on the non-existence of English associationism. J Hist Behav Sci 1(2):123–130

Ferg S (1981) Two early works by David Hartley. J Hist Philos 19(2):173–189

Gertler B (2011) Self-knowledge. Routledge, New York

Gaukroger S (2009) The role of natural philosophy in the development of Locke's empiricism. Br J Hist Philos 17(1):55–83

Gray R (1978) Hobbes' system and his early philosophical views. J Hist Ideas 39(2):199–215

Hartley D (1834) Observations on man, his fame, his duty, and his expectations. Thomas Tegg and Sons, London

Hobbes T (1994) Leviathan (Curley, ed. & notes). Hackett, Indianapolis

Hobbes T (2008) Human nature and De Corpore Politico (Gaskin, ed. & notes). Oxford University Press, Oxford

Hume D (1978) A treatise of human nature (Nidditch, text rev. & notes). Clarendon Press, Oxford

James W (1952) The principles of psychology. Encyclopædia Britannica, Chicago

Klein DB (1970) A history of scientific psychology: its origins and philosophical backgrounds. Basic Books, New York

Locke J (1979) An essay concerning human understanding (Nidditch, ed.). Clarendon Press, Oxford

Loeb LE (2006) Psychology, epistemology, and skepticism in Hume's argument about induction. Synthese 152(3):321–338

Mandler G (2007) A history of modern experimental psychology. MIT Press, Cambridge

McRae R (1965) "Idea" as a philosophical term in the seventeenth century. J Hist Ideas 26(2):175–190

Nathanson SL (1973) Locke's theory of ideas. J Hist Philos 11(1):29–42

Nagel E (1970) Issues in the logic of reductive explanations. In: Keifer H and Munitz M (eds) Mind, science, and history. SUNY Press, Albany, pp 117–137

Pillsbury WB (1929) The history of psychology. W. W. Norton & Company, New York

Robinson DN (1986) An intellectual history of psychology. University of Wisconsin Press, Madison

Rogers GAJ (1978) Locke's essay and Newton's Principia. J Hist Ideas 39(2):217–232

Rogers GAJ (1996) Science and British philosophy: Boyle and Newton. In Brown S (ed) British philosophy and the age of enlightenment. Routledge, New York

Ryle G (2000) The concept of mind. Penguin Books, London

Smith CUM (1987) David Hartley's Newtonian neuropsychology. J Hist Behav Sci 23:123–136

Soles DE (1985) Locke's empiricism and the postulation of unobservables. J Hist Philos 23(3):339–369

Titchener EB (1926) A text-book of psychology. The Macmillan Company, New York

Warren HC (1967) A history of the association psychology. Charles Scribner's Sons, New York

Yolton JW (1963) The concept of experience in Locke and Hume. J Hist Philos 1(1):53–71

Chapter 2
Mature British Associationism

2.1 David Hartley

If the mansion of the human psyche was sketched by Hobbes, Locke and Hume, the detailed, architectural blueprint was drawn during the mature phase of associationist thought, first by a famous physician following in Locke's footsteps. I am of course referring to Hartley (1705–1757), a contemporary of Hume who aimed to fuse the perspectives of Newton[1] and Locke into one comprehensive doctrine that was part neurophysiological and part associationistic.[2] As Hartley says in the first chapter of his *Observations on Man, His Frame, His Duty, and His Expectations*, his

> chief design in the following chapter is briefly to explain, establish, and apply the doctrines of vibrations and association. The first of these doctrines is taken from the hints concerning the performance of sensation and motion, which Sir Isaac Newton has given at the end of his Principia, and in the Questions annexed to his Optics; the last, from what Mr. Locke, and other ingenious persons since his time, have delivered concerning the influence of association over our opinions and affections … (Hartley 1834, p. 4, italics removed).

Hartley, as Boring puts it, "was the founder of associationism" and whoever discovered associationism "there is not the least doubt that Hartley prepared it for its *ism*." (Boring 1950, p. 194). The main statement of that "ism" is found in Hartley's *Observations* first published in 1749.[3]

---

[1] As Mischel notes, Hartley falls into the tradition of "*speculative* Newtonian science common in that century" (Mischel 1966, p. 126, emphasis added), not *experimental* Newtonianism. The serious attempt to bring British associationism into the laboratory, as we shall see, was first made more than a century later.

[2] Hartley seems to have been unsure as to whether the neurophysiological side was "necessary to the total theory" or whether it was merely a "complementary and parallel explanation" (Oberg 1976, p. 442).

[3] It has been argued that two earlier works, published anonymously in 1741 and 1747 and advancing the associationist theory of psychology in preliminary form, should also be attributed to Hartley (Ferg 1981). We will set this aside and here focus on the *Observations*.

C. Beenfeldt, *The Philosophical Background and Scientific Legacy of E. B. Titchener's Psychology*, SpringerBriefs in Philosophy, DOI: 10.1007/978-3-319-00242-2_2,

Again, we find sensationism. Sensations are those "internal feelings of the mind" that arise from the stimulation by external objects of the appropriate parts of our body (Hartley 1834, p. 1) and all other mental content may be divided into simple and complex ideas. Simple ideas are ideas that "resemble sensations" (Hartley 1834, p. 1), the "copies and offsprings of the impressions made on the eye and ear" (Hartley 1834, p. 36), the elementary building blocks out of which everything else is constructed. Complex ideas, also called intellectual ideas, are those compounded units that constitute the rest of our mental life (Hartley 1834, p. 1). In Hartley's own words, "ideas of sensation … [may] be called simple ideas, in respect of the intellectual ones which are formed from them, and of whose essence it is to be complex" (Hartley 1834, p. 36, italics removed). This notion is closely related to the doctrine of elementism because, for Hartley, the ideas of sensation "are the elements of which all the rest are compounded" (Hartley 1834, p. 1, italics removed).

Assuming, then, that sensations are the building blocks of mind, what is the associative force that binds them together? Unlike Hume, Hartley recognizes repeated contiguity in time as the sole basis for association (Warren 1967, p. 55). As he puts it, any

> sensation A, B, C, &c., by being associated with one another a sufficient Number of Times, get such a Power over the corresponding Ideas a, b, c, &c. that any one of the Sensations A, when impressed alone, shall be able to excite in the Mind, b, c, &c. the Ideas of the rest (Hartley 1834, p. 41, italics removed).

Accordingly, if we want to account for our idea of a horse, say, we observe that the "particular ideas of the head, neck, body, legs, and tail, peculiar to this animal, stuck to each other in the fancy, from frequent joint impressions" (Hartley 1834, p. 45). Memory is simply the recurrence of "traces of sensations and ideas" (Hartley 1834, p. 235), and so essentially is imagination, although it has less vivid content. Emotions are "aggregates of simple ideas united by association" (Hartley 1834, p. 231), words and phrases also "excite Ideas in us by Association" (Hartley 1834, p. 169, italics removed) and dreams occur when the associative bond is weakened at night, and "various parcels of visible ideas, not joined in nature, start up together in the fancy … [making us] often see monsters, chimeras, and combinations, which have never been actually presented" (Hartley 1834, p. 45).

What Hartley offers us here is a highly reductionistic project. The elementary ideas gather, combine and dissolve their compounded relations by a simple mechanism, like so many billiard balls on a table.[4] Indeed, like Hobbes before him, Hartley is not merely a decompositional reductionist but also a mechanistic reductionist. Mechanism, in this context, is a specific sub-type of reductionism according to which the system in question is governed by the mechanical law of push-and-pull (or something analogous thereto). Hobbes described sense perception in terms of the complex chain reactions of internal pressure and counter pressure in our nerves, membranes, brain and heart. This was a thoroughly mechanical account that, less than a century later, might have been used to describe the workings of Newcomen's

[4] This metaphor has also been applied to Locke's philosophy of mind (Smith 1987, p. 125).

steam engine. Hume and Locke were less enthralled than Hobbes by the project of reducing everything to the push-and-pull of matter in motion, but in Hartley we find this perspective resurfacing. "Mechanism" as he put it in a letter to Lister in 1739, is "the true Hypothesis," "I am mechanical, [and] for this very Reason it is that the Flame [which] I see approaching, raises in me Apprehension [and] a Train of internal Sensations, the consequence of [which] is (according to the preceding Associations) that my muscles contract [and] carry me away ..." (quoted in Webb 1988, p. 205).

The reductionistic aim is to "analyze all that vast variety of complex ideas, which pass under the name of ideas of reflection, and intellectual ideas, into their compounding parts, i.e. into the simple ideas of sensation, of which they consist" (Hartley 1834, p. 48). In short, the goal is to explain our entire mental life in terms of the interchange of atomic experiences. Just as the law of gravity explains the movement of planets and moons in our solar system, so does the principle of association explain human mental life and all its products, including "the power of habit, custom, example, education, authority, party-prejudice, the manner of learning the manual and liberal arts, &c" (Hartley 1834, pp. 41–42): it all comes down to a single, ultimate and ruling causal factor, namely the association-inducing contiguity of experiences.[5]

2.2 James Mill

We now move to the Mills, father and son. We will first consider Mill (1773–1836), whose work has been characterized as both the culminating effort of associationism (Boring 1950, p. 203) and the beginning of modern English psychology (Pillsbury 1929, p. 130). The main contribution by Mill was his *Analysis of the Phenomena of the Human Mind* (1829), which can be understood as an attempt to extend Hartley's views on the association of ideas, *minus* all the physiological considerations (Mischel 1966, p. 130). If any work qualifies as a textbook in pure associationism, this is a very good candidate indeed. In Brett's estimate, Mill was nothing less than "the accepted oracle of associationism" (Brett 1973, p. 41).

According to the form of sensationism advanced by Mill, our basic mental content is "feeling," an item that then divides into two classes: sensation and ideas (Mill 1869, p. 52). The distinction is straightforward. Sensations exist "when the object of sense is present" and "ideas exist after the object of sense has ceased to be present" (Mill 1869, p. 52). As Mill explains, when "I have seen the sun, and by shutting my eyes see him no longer, I can still think of him" (Mill 1869, pp. 51–52). That thought is then "a copy, an image, of the sensation; sometimes, a representation, or trace, of the sensation" (Mill 1869, p. 52). Or, again, we can think of colors in the dark. When we do so, we have a feeling "which is not the same with the sensation, but which we

[5] It must be remembered that, while Hartley was pulled in two opposite directions by his commitments to Newtonian science and associationist psychology on the one hand, and clerical training as a Christian philosopher on the other hand (Oberg 1976, p. 442), he "was close to a materialist explanation of man, and his critics thought he had reached it" (Oberg 1976, p. 444).

contemplate as a copy of the sensation, an image of it," a copy that may then later be redeployed in a train of thoughts (Mill 1869, p. 54). Sensations come to us from our five senses as well as from our muscle sense and the alimentary canal. Indeed, in Mill's system we cannot reach beyond our own sensations, and so in "using the names, tree, horse, man, the names of what I call objects, I am referring, and can be referring, only to my own sensations" (Mill 1869, p. 93).

In the same vein, Mill advances a position on the primary-secondary qualities distinction in which everything, (in the final analysis, anyway) refers to our own sensations, and to quite unsophisticated ones at that. Consider Mill's example of our idea of gold. To begin with, the idea divides into ideas of several sensations, namely "color, hardness, extension, weight" (Mill 1869, p. 91). Let us focus on weight, which "appears so perfectly simple, that he is a good metaphysician who can trace its composition" (Mill 1869, p. 91). Fortunately, Mill is just such a metaphysician and so we learn that this notion involves the idea of resistance, which, in turn, involves "the feeling attendant upon the contraction of muscles" (Mill 1869, p. 91). This further involves not simply the idea of resistance, but the idea of resistance in a particular direction, and that extension, place, and motion (Mill 1869, p. 91). Extension, to follow this analysis one final step, is ultimately derived from "muscular feelings" (Mill 1869, p. 92).

How are the components of mental life, the ideas and their constituent sensations, organized? Put simply, they are organized according to the laws of association, of course. To be more specific, according to Mill, we can distill these laws into two principles: succession and synchronicity. Succession pertains to the sequential order of the elements, such as when the thought of a horse might lead one to the idea of its owner and then to the idea of the owner's profession. Synchronicity pertains to the grouping of the elements at a given point in the sequence, such as when hardness, a certain shape, and a characteristic shade of gray go together in the tightly knit bundle that we call a stone. Armed with this Mill proceeds to reduce Hume's principles of association, the principle of contiguity, of causation, and of resemblance—as follows: contiguity in space is simply synchronous order and contiguity in time is successive order; causation is successive order; and resemblance is a matter of frequency.

Whatever the principles of association are eventually refined to be, it is clear that Mill's system reduces the complex mental phenomena (e.g. our idea of gold) to the most basic of human sensory experiences. Thus, our mental life can be accounted for as a mass of atomic sensations succeeding each other and grouping together associationistically, like so many ideational pearls on a string or mental balls in a game of pétanque. The laws of the mind that govern the composition of the simple into the complex are "analogous to mechanical … laws," as Mill's son explained. In short, Mill's *Analysis*, propounds

> an extreme form of associationism, in which all man's complex intellectual and moral states are analyzed into combinations of "sensations, ideas, and trains of ideas" (Hearnshaw 1964, p. 2).

The system is a *mental physics*.

2.3 John Stuart Mill

Consider now Mill the son, Mill (1806–1873), whose thoughts on associationism were mainly expressed in his published annotations to his father's *Analysis*, as well as in his own *Logic* from 1843,[6] in particular the first part of Book VI, in the section entitled "On the logic of the moral sciences."[7] The aim of John Stuart's work in psychology was to amend his father's account: to retain the elementist-reductionist-sensationist core assumptions while defusing the criticism that this view was facing at the time (Warren 1967, p. 95). As Randall stated,

> [h]e took over the principles of Association, and the assumptions bout up with it, that experience is equivalent to isolated feelings, and that all relations in it are forms of association (Randall 1965, p. 61).

Let us here focus on the single most salient item, namely John Stuart Mill's amended account of the construction of complex mental elements from simple ones.

In Mill's view, the process works by a relatively straightforward mechanism that is analogous to the operation of Newtonian mechanics on physical bodies. Here, a complex whole simply equals the addition of its parts, such as when we consider a medieval dry stack stone wall as a structure simply composed of individual stones. As Mill's sees it, however, simple ideas sometimes *generate* complex ideas rather than merely compose them. He calls this *mental chemistry* (Mill 2006, pp. 853–854). As he explains, "the laws of mental phenomena are sometimes analogous to mechanical, but sometimes also to chemical laws" (Mill 2006, p. 853).

Thus, while James Mill posited only mechanical laws of the mental, John Stuart Mill now adds the further claim that there are also distinctively chemical laws at work in this realm. Specifically, when many impressions or ideas are operating in the mind together there "sometimes takes place a process of a similar kind to chemical combination" (Mill 2006, p. 853). When impressions have been "so often experienced in conjunction, that each of them calls up readily and instantaneously the ideas of the whole group, those ideas sometimes melt and coalesce into one another, and appear" not as "several ideas, but [as] one" (Mill 2006, p. 853). Whereas the idea of an orange can be simply decomposed, at least in the first instance, into the individual ideas of a certain color, a certain form, a certain taste, and a certain smell etc., the idea of extension (which Mill in his son's estimation had conclusively shown to originate from the muscular sensation of resistance) cannot be similarly decomposed. This is, rather, a case of chemical change,

---

[6] In full, *A System of Logic Ratiocinative and Inductive: Being a Connected View of the Principles of Evidence and the Methods of Scientific Investigation* (Mill 2006).

[7] Also in *An Examination of the Philosophy of Sir William Hamilton* (Mill 1865).

where a complex item is a product of generation and not of mere composition (Mill 2006, pp. 848–860).[8]

It is doubtful whether this represents a real improvement on Mill's mental physics and not simply the equivalent of a Ptolemaic epicycle attached to the system to attenuate its apparent implausibility. In any event, Stuart Mill's theory leaves the essential associationistic assumptions untouched. In mental chemistry the complex wholes in human mental life are no longer viewed as straightforward compositions of the simple elements (sensations) and their derivatives (ideas), brought about by the process of association. Rather, the existence of intermediate simples, of ideas that have melted and coalesced into one another are posited. Yet, to introduce a fusion process rather than a mere collection process does not change the fact that elemental simples alone constitute the only input to the system (Randall 1965, p. 62). Further, as science has progressed since Mill's time, we now understand that chemical changes are fully in line with, and explicable in terms of, the laws of physics. Chemical fusion, indeed, *can* be understood in terms of straightforward physical change happening at a sufficiently fine-grained microphysical level.[9]

Viewed from the distance of more than a century, mental physics and mental chemistry really do not stand very far apart. Sounding very much like his father, Mill prefixed his discussion of mental chemistry by stating that the

> subject, then, of Psychology, is the uniformities of succession, the laws, whether ultimate or derivative, according to which one mental state succeeds another, is cause by, or at least, is cause to follow, another (Mill 2006, p. 852).

2.4 A Physical Science of the Mental

To summarize our findings so far, the central aim of associationism was to provide what was understood to be a scientific account of the human mind—one explicitly modeled on the advances in the physical sciences in general and on the great achievements of Newtonian physics in particular. This project was seen to involve (1) a reductive decomposition of human mental life into (2) elements that ultimately are (3) sensationistic in nature, and the aim of enquiry was (4) the discovery of the laws of association for human psychology:

1. "[S]imple Ideas" are, according to Locke, "the Material of all our Knowledge," and when "the Understanding is once stored with these simple Ideas,

---

[8] Strictly speaking, this amounts to an attenuated form of decompositional reductionism since the mid-level fusion events stop any decompositional reduction back to the original elements. Recognizing this caveat, however, we will continue to speak of John Stuart Mill's position, like that of his father, as a form of associationism fundamentally committed to both elementism and to decompositional reductionism. Again, the aim of the younger Mill was to answer criticism of associationism—*not* to abandon or to substantially alter the basic theory.

[9] Mill's own parallel example of the compound water as an intermediary simple makes this clear (Mill 2010).

it has the Power to repeat, compare, and unite them even to an almost infinite variety, and so can make at Pleasure new complex Ideas" (Locke 1979, p. 119). The aim of psychology is thus, in Hartley's judgment, to "analyze all that vast variety of complex ideas, which pass under the name of ideas of reflection, and intellectual ideas ... into the simple ideas of sensation, of which they consist" (Hartley 1834, p. 48). Imagination, memory, and dreams, for example, can thus be understood as "decaying sense" as Hobbes thought (Hobbes 1994, p. 8).[10]

2. The ideas of sensation, in Hartley words, "are the elements of which all the rest are compounded" (Hartley 1834, p. 1, italics removed), and thus, ideas of sensation may be termed simple and, intellectual ideas, complex (Hartley 1834, p. 36). The distinction is straightforward: in simple form, as Hume told us, mental items "admit of no distinction or separation" but, in complex form, they do admit of separation into constituent parts (Hume 1978, p. 2).

3. Our senses, in Locke's words, "convey into the Mind" and thus "we come by those Ideas, we have of Yellow, White, Heat, Cold, Soft, Hard, Bitter, Sweet, and all those which we call sensible qualities" (Locke 1979 p. 105, italics removed). As Hartley put it, "internal feelings of the mind" arise from the stimulation by external objects of the appropriate parts of our body (Hartley 1834, p. 1), and simple ideas are the copies and "offsprings of the impressions made on the eye and ear" (Hartley 1834, p. 36) out of which the human mental life is ultimately composed. Strictly speaking, in Hume's view, we do not really perceive our own body when "we regard our limbs and members." Rather, we absorb "certain impressions, which enter by the senses" (Hume 1978, p. 191). Most radically, one might claim along with James Mill that by "using the names, tree, horse, man, the names of what I call objects", I "am referring, and can be referring, only to my own sensations" (Mill 1869, p. 93).

4. The laws of association answer the question of *how*—given elementism, decompositional reductionism, and sensationism—the elements unite into complex structures. Is this process predictable and describable by universal interaction laws, such as the law of gravity, or is it rather random and chaotic? As a school thinkers approaching psychology with Galileo, Boyle,[11] and Newton[12] as their models of achievement,[13] the British associationists had a firm commitment to the first option. They differed much on the exact formulation of the laws of association, but they all agreed that it *is* universal and law-like. Hume spoke of "universal principles" concerning an attraction in the mental world as powerful as the force

[10] For Hartley and Mill, explaining mental data, "does not differ essentially from explaining the properties of a mineral," as one commentator put it (Mischel 1966, p. 144).

[11] Boyle was the great British advocate of the atomic view in natural philosophy. His influence on the Royal Society and the intellectual community in general was vast (Rogers 1996).

[12] Newton and Locke became friends, as the story goes, in late 1689 or 1690 (plausibly) meeting first at the salon of the Earl of Pembroke (Rogers 1978, p. 231).

[13] Not forgetting, of course, that philosophers are a quarrelsome bunch. For example, in *Leviathan and the Air-Pump* Shapin and Schaffer (1989) offer a detailed account of Hobbes's conflict with Robert Boyle and the Royal Society.

of gravity in the natural world (Hume 1978, p. 10), and Hartley speculated about reverse-engineering the process, holding it to be of the "utmost consequence to morality and religion, that the affections and passions should be analyzed into their simple compounding parts, by reversing the steps of the associations which concur to form them" (Hartley 1834, p. 52).

The final aim was to furnish a physical science of the mental, in which the physical corpuscles became the sensory elements and gravitational attraction became mental association. In other words, one can understand

> the development of English association psychology during the [eighteenth and nineteenth centuries] as an attempt to construct a "science of the mind" modeled on the physical sciences. Writers of this tradition wanted to explain how the emotions "move" to action without invoking mysterious powers or agencies, just as physicists were explaining motions without appealing to occult qualities (Mischel 1966, p. 123).

Gary Hatfield has summed up the associationist "explanatory strategy" as the attempt to "discern or posit elements of consciousness and then to show how the laws of association, operating on such elements, can explain mental abilities and mental phenomena more generally" (Hatfield 2003, p. 95). The associationist approach was misguided. Human psychology is not a vast system of sensory atoms interacting according to quasi-gravitational laws of attraction for the mental. Rather than good scientific psychology, this is an example of how impressive scientific breakthroughs in one domain can be carried—speculatively—far beyond their proper bounds.

In his excellent discussion of John Stuart Mill, Randall contends that traditional British empiricism died with Mill (Randall 1965, p. 60). As far as the history of philosophy is concerned, this may well be true. However, we will now see that the spirit of traditional empiricism survived in the form of a highly influential system of psychology that aimed to *fuse* the deep heritage of associationist psychology with what was quickly becoming the prevailing form of scientific investigation: experimental laboratory work. We will also see how that system collapsed, by dint of what can only be termed a great misattribution of guilt, thereby bringing introspection into general disrepute in psychology for most of the twentieth century.

References

Boring EG (1950) A history of experimental psychology. Appleton-Century-Crofts, New York

Brett GS (1973) Associationism and "act" psychology: a historical retrospect. In: Murchinson C (ed) Classics in psychology. Arno Press, New York

Ferg S (1981) Two early works by David Hartley. J Hist Philos 19(2):173–189

Hartley D (1834) Observations on man, his fame, his duty, and his expectations. Thomas Tegg and Sons, London

Hatfield G (2003) Psychology: old and new. In: Baldwin T (ed) The Cambridge history of philosophy 1870–1945. Cambridge University Press, Cambridge

Hearnshaw LS (1964) A short history of British psychology 1850–1940. Methuen & Co, London

Hobbes T (1994) Leviathan (Curley, ed. & notes). Hackett, Indiannopolis

Hume D (1978) A treatise of human nature (Nidditch, text rev. & notes). Clarendon Press, Oxford

Locke J (1979) An essay concerning human understanding (Nidditched, ed). Clarendon Press, Oxford
Mill JS (1865) An examination of Sir William Hamilton's philosophy. Longmans, Green, and Co, London
Mill J (1869) Analysis of the phenomena of the human mind (Mill JS, ed & additional notes). Longmans Green Reader and Dyer, London
Mill JS (2006) Collected works of John Stuart Mill, vol VIII (Robson, text ed). Liberty Fund, Indianapolis
Mill JS (2010) Bain's psychology. The online library of liberty. Web. April 10. http://oll.libertyfund.org/?option=com_staticxt&staticfile=show.php%3Ftitle=248&chapter=21776&layout=html&Itemid=27
Mischel T (1966) Emotion and motivation in the development of English psychology: Hartley D, Mill J, Bain A. J Hist Behav Sci 2(2):123–144
Oberg BB (1976) David Hartley and the association of ideas. J Hist Ideas 37(3):441–454
Pillsbury WB (1929) The history of psychology. W. W. Norton & Company, New York
Randall JH (1965) John Stuart Mill and the working-out of empiricism. J Hist Ideas 26(1):59–88
Rogers GAJ (1978) Locke's essay and Newton's Principia. J Hist Ideas 39(2):217–232
Rogers GAJ (1996) Science and British philosophy: Boyle and Newton. In: Brown S (ed) British philosophy and the age of enlightenment. Routledge, New York
Shapin S, Schaffer S (1989) Leviathan and the air-pump. Princeton University Press, Princeton
Smith CUM (1987) David Hartley's Newtonian neuropsychology. J Hist Behav Sci 23(2):123–136
Warren HC (1967) A history of the association psychology. Charles Scribner's Sons, New York
Webb ME (1988) A new history of Hartley's Observations on Man. J Hist Behav Sci 24(2):202–211

Part II
The System of Introspectionism

Chapter 3
Wundt and Titchener

3.1 Wundt and the Beginning of Modern Psychology

The year 1879 is generally regarded as seminal in the history of psychology; it is widely agreed that this marks the official beginning of modern psychology.[1] It was the year that Wilhelm Maximilian Wundt (1832–1920) founded, at the University of Leipzig in the then recently unified German state, what has come to be regarded as the world's first laboratory of psychology.[2] Wundt received his medical training at the time when the great physiologists, Hermann von Helmholtz among them, were establishing the field of physiology as an independent experimental science, distinct from anatomy (Danziger 1990, pp. 24–25).

Wundt sought to adopt the experimental method from physiology and he "obviously hoped [that] this recent rather impressive success [in physiology] might serve as the model for the transformation of another field, namely psychology" (Danziger 1990, p. 25). In Wundt's famous laboratory we find the important

[1] Despite widespread agreement on this point, the marking of this date as the beginning of modern psychology is not met with universal assent. Hatfield has argued that it "obscures the disciplinary and theoretical continuity of the new experimental psychology with a previous, natural philosophical psychology. And it goes together with a story of rapid antagonism between philosophy and psychology at century's turn, which itself seriously misrepresents the state of play between philosophers and psychologists at the time" (Hatfield 2002, p. 209). The classic discussion of this issue is Boring's (1965) paper "On the subjectivity of important historical dates: Leipzig 1879," where he concludes that there is "a very considerable element of subjectivity in the establishment of this date" (Boring 1965, p. 6). For our present purposes, we shall not take a position on this matter. However, it is worthwhile to note that one purpose of the present chapter is to make salient the profound philosophical continuity that exists between the associationistic thought of Hobbes, Locke, Hume, Hartley, and the Mills on the one hand, and the subsequent introspectionist body of thought in experimental psychology on the other hand.

[2] On the role of the German academic science culture in the development of scientific psychology, see Dobson and Bruce (1972).

C. Beenfeldt, *The Philosophical Background and Scientific Legacy of E. B. Titchener's Psychology*, SpringerBriefs in Philosophy, DOI: 10.1007/978-3-319-00242-2_3,

division of labor in psychology between the experimenter (*Beobachter*) and the experimental subject (*Versuchsperson*), which

> was an extremely important development with the most profound implications for the nature of psychological research. The division of labor that was spontaneously adopted in Wundt's laboratory was none other than the well-known division between the roles of "experimenter" and "subject" in psychological experiments (Danziger 1990, p. 30).

Wundt characterized his own approach as "physiological psychology" and his monographic contributions include the *Grundzüge der physiologischen Psychologie*, a work still awaiting comprehensive scholarly treatment. Even less well-studied is his immense ten-volume *Völkerpsychologie: Eine Untersuchung der Entwicklungsgesetze von Sprache, Mythus und Sitte* in which he investigates the higher reaches of human mental life, including cultural products such as art, myth and customs. We would today recognize the work as sharing investigative object domains with areas such as history, philology, linguistic, ethnology and anthropology, here analyzed with a view to drawing inferences about the psychological processes involved in their production (Greenwood 2003, p. 75). This non-experimental side of Wundt's work, however, was, in the context of scientific psychology, all but ignored even while the experimental side attracted both attention and emulation.

During his long career, Wundt taught as many as 24,000 students and supervised nearly 200 doctoral dissertations; indeed, the list of his doctoral students reads like an index to a history of modern psychology (Bringmann et al. 1975, p. 294). Wundt is generally regarded as the founder of experimental psychology, by which, in Boring's words, "we mean both that he promoted the idea of psychology as an independent science and that he is the senior among 'psychologists'" (Boring 1950, p. 316).

Wundt's most famous student was Edward Bradford Titchener (1867–1927). Titchener was born in 1867 in Chichester, England, about 70 miles south of London. He went to Oxford in 1885 and was a member of Brasenose College, first as a philosophy and classics scholar, then (in his fifth year) as a research student of physiology (Boring 1927, p. 490). Completing his studies at Oxford, Titchener began his research in experimental psychology under the tutelage of Wundt in Leipzig. Having received his Doctor of Philosophy degree in 1892, Titchener served for a few months as an extension lecturer in biology at Oxford and then relocated to Cornell University (Boring 1927, p. 493). He spent the last 35 years of his career at Cornell University, where he exercised considerable influence upon contemporary thought in psychology, especially in America.

3.2 The Wundt-Titchener Relationship

Much debate in history of psychology circles has centered on the question of the precise relationship between Wundt and Titchener. On one side there is the *traditional* account, advanced most notably by the prominent Wundt students, Titchener and Oswald Külpe—and, in turn, by Titchener's own student, Edwin G. Boring, who carved the traditional view into the psychological record with his highly influential *A History of Experimental Psychology*.[3] By his very "definition of psychology," according to this account, Wundt made "introspection for the time being the primary method of the psychological laboratory" (Boring 1950, p. 328)—a state of affairs that lasted "until behaviorism came into vogue in America (ca. 1913)" (Boring 1950, p. 332). Based on his reading of the Wundtian oeuvre, Boring concludes that "there was a body of opinion which was in general shared by Wundt, by Külpe before he left Leipzig, by G. E. Müller at Göttingen, by Titchener at Cornell and by many other less important 'introspectionists' who accepted the leadership of these men" (Boring 1953, p. 171). Titchener, on the traditional account, is viewed as the leading introspectionist in America around the turn of the twentieth century precisely because he "outwundted Wundt" (Boring 1950, p. 386)—precisely, that is, because he excised the inconsistent elements in Wundt's work and represented introspectionism in *pure form* (Boring 1950, p. 386).[4]

According to the opposing *modern account* of *Wundt*, advanced by a number of late twentieth century historians of psychology (e.g. Blumenthal 1979, Costall 2006, Danziger 1979, 1980 and Leahey 1981), the traditional image of Wundt is a distortion.[5] In *The Cognitive Revolution in Psychology*, Bernard Baars sums up the significance of the new account by asserting that the conception of Wundt as "an introspectionist, an associationist, a dualist, and a believer in 'mental chemistry'" now "appears almost entirely false" (Baars 1986, p. 30).

The scholarly criticism of the traditional account centers on the historical work of Boring who, it is claimed, seriously misunderstood Wundt. Boring, in turn, acquired his own mistaken understanding from Titchener, his teacher[6] and the man to whom he dedicated[7] his famous work on the history of psychology.[8] The consequences of this error, according to the modern account, have been so extensive that, as Costall describes in a recent contribution to *Consciousness and Cognition*, the *myth* of Wundt as the arch-introspectionist persists in the current work of Adams, Blackmore, Dennett, Güzeldere, Hooker, Kihlstrom, Hooker, Leahey,

[3] From time to time, the traditional account also slips into the writings of other twentieth century historians of psychology. Robert Watson, for example, claimed that Titchener had an "unshakable allegiance" to Wundt (Watson 1965, p. 131). We also see it in Hatfield's claim that Titchener was "pursuing the Wundtian project of resolving mental life into its elements" (2003, pp. 103).

[4] This was also how Titchener himself viewed it. He saw himself as extending the domain of experimental methodology beyond the self-limiting boundaries that Wundt had placed upon his own investigations (see Titchener 1920, p. 502).

[5] Already in 1887, however, we see a growing recognition on the part of Bain, of Wundt's distinctive position regarding the "insufficiency or shortcoming of the principles of Association" (Bain 1887, p. 174).

[6] And in whose shadow he walked as a graduate student (Boring 1967, p. 315).

[7] The inscription simply reads "To Edward Bradford Titchener."

[8] The problem is acute because "Titchener and Boring were key figures in carrying the burden of explanation of Wundt's work" (Anderson 1975, p. 385).

Lundin, Rosch, Rosenthal, Thompson, Varela, Vermersch and Weidman (Costall 2006, p. 646). These thinkers have all been misled by Titchener who sought to create the impression that his own introspectionist system, through Wundt, had deep roots extending back to the very founding soil of experimental psychology. He accomplished this by systematically reinterpreting Wundt's theories "to remove or trivialize their basic incompatibility with the classical British tradition in psychology" as well as with Titchener's own philosophy of science (Danziger 1979, p. 217; Danziger 1980, pp. 244 and 246).

For our present purposes, *this* debate can be bracketed. Let us consider the strongest suggested positions at each end of the spectrum, namely (1) that Wundt was the true forbearer of introspectionist thought, or (2) that Wundt's public image as an introspectionist was in fact a perverse misrepresentation by Titchener, Külpe and Boring. In either case, and in all instances in-between, the *perceived image* of Wundt was that of an introspectionist whose views on experimental psychology were proximal to those of Titchener, his de facto spokesman in the English-speaking world. It was this perceived state of affairs at that time, and subsequently, that is causally relevant with respect to the theory of introspection in psychology that was to dominate.

When Watson launched behaviorism in direct opposition to "introspective psychology" (Watson 1966, p. 3, italics removed), he was reacting to a research program that was widespread in America at the time.[9] Wundt was, rightly or wrongly, taken to be the originator of this approach, while Titchener was its undisputed leader and prime Anglo-American representative (Boring 1937, p. 470; Schwitzgebel 2004, p. 60). Titchener can, in this sense at least, be called the premier introspectionist in the history of psychology, the pinnacle of the—golden or gilded—age of introspection.[10] We will now examine his perilously influential system of psychology.

[9] Watson was, in fact, casting an extremely wide net in finding opponents to his own position. In *Behaviorism*, he takes exponents of "the older psychology … called introspective psychology" to include not only Wundt, Külpe, and Titchener—but *also* James, Angell, Judd and McDougall (Watson 1966, p. 3, italics removed). What unifies these very different thinkers, in Watson's mind, is that they all claim that consciousness "is the subject matter of psychology" (Watson 1966, p. 3, italics removed).

[10] In *The Disappearance of Introspection*, William Lyons characterizes the period from the seventeenth century to the first decade of the twentieth century as "the golden age of introspection" (Lyons 1986, p. 2). As Lyons further states, "[i]ntrospection in its classical form (or forms) may be said to have reached its zenith and nadir at the same time in the school of Titchener in the United States" (Lyons 1986, p. 21). The zenith also represented, in a different sense, the nadir because the (purported) practice of introspection had "become highly elaborated by the time and, to the growing number of outsiders, bizarre" (Lyons 1986, p. 21). We shall uncover the cause of this as we proceed. As we shall also see, this period was more of a gilded than a golden age of introspection.

References

Anderson RJ (1975) The untranslated content of Wundt's Grundzüge Der Physiologischen Psychologie. J Hist Behav Sci 11(4):381–386
Baars BJ (1986) The cognitive revolution in psychology. Guilford, New York
Bain A (1887) On 'Association' - controversies. Mind 12(46):161–182
Blumenthal AL (1979) The founding father we never knew. Contemp Psychol 24(7):547–550
Boring EG (1927) Edward Bradford Titchener: 1867–1927. Am J Psychol 38(4):489–506
Boring EG (1937) Titchener and the existential. Am J Psychol 50(1/4):470–483
Boring EG (1950) A history of experimental psychology. Appleton-Century-Crofts, New York
Boring EG (1953) A history of introspection. Psychol Bull 50(3):169–189
Boring EG (1965) On the subjectivity of important historical dates: Leipzig 1879. J Hist Behav Sci 1(1):5–9
Boring EG (1967) Titchener's experimentalists. J Hist Behav Sci 3(4)
Bringmann WG, Balance WDG, Evans RB (1975) Wilhelm Wundt 1832–1920: a brief biographical sketch. J Hist Behav Sci 11(3):287–297
Costall A (2006) 'Introspectionism' and the mythical origins of scientific psychology. Conscious Cogn 15(4):634–654
Danziger K (1979) The positivist repudiation of Wundt. J Hist Behav Sci 15(3):205–230
Danziger K (1980) The history of introspection reconsidered. J Hist Behav Sci 16(3):241–262
Danziger K (1990) Constructing the subject: historical origins of psychological research. Cambridge University Press, Cambridge
Dobson V, Bruce D (1972) The German university and the development of experimental psychology. J Hist Behav Sci 8(2):204–207
Greenwood JD (2003) Wundt, Völkerpsychologie, and experimental social psychology. Hist Psychol 6(1):70–88
Hatfield G (2002) Psychology, philosophy, and cognitive science: reflections on the history and philosophy of experimental psychology. Mind Lang 17(3):207–232
Hatfield G (2003) Psychology: old and new. In: Baldwin (ed) The Cambridge history of philosophy 1870–1945. Cambridge University Press, Cambridge
Leahey TH (1981) The mistaken mirror: on Wundt's and Titchener's psychologies. J Hist Behav Sci 17(2):273–282
Lyons W (1986) The disappearance of introspection. MIT Press, Cambridge
Schwitzgebel E (2004) Introspective training apprehensively defended: reflections on Titchener's lab manual. J Conscious Stud 11(7–8):58–76
Titchener EB (1920) Wilhelm Wundt, 1832–1920. Science, New Series 52(1352):500–502
Watson RI (1965) The historical background for national trends in psychology: United States. J Hist Behav Sci 1(2):130–138
Watson JB (1966) Behaviorism. University of Chicago Press, Chicago

Chapter 4
Titchener's System of Psychology

4.1 A Structural Psychology

Titchener, it has been observed, "always disliked labels as pinning one down" (Boring 1927, p. 497). He did not recognize himself as an "introspectionist." Indeed, so extreme was his dislike of labels that he, characteristically, did not refer to "his [own] school by any other words than 'we'" (Boring 1927, p. 497). Titchener did find the term "structuralism" for his own system somewhat useful, because it stressed the contrast between his own approach and that of functional psychology, but he was never happy with this nomenclature (Evans 1984, pp. 120–122).

The point to make about Titchener's psychology,[1] then, is that despite his great fame (and, later, notoriety) as the grand introspectionist in Anglo-American psychology, he would in the first place have been more inclined to recognize himself as advancing a *structural psychology* (Titchener 1898 and 1899). In his 1898 paper, "The Postulates of a Structural Psychology," in fact, he draws a parallel between the work performed in experimental psychology and the work performed in morphology, the branch of biology dealing with the structure of organisms. He maintains that the "primary aim of the experimental psychologist" is, as in morphology, to "analyze the structure of mind; to ravel out the elemental processes from the tangle of consciousness" (Titchener 1898, p. 450).

More broadly, he contends that scientific psychology must fall into line with the remaining sciences, all of which "find their source in analysis" (Titchener 1896,

[1] We will here be very critical of Titchener's psychology and experimental methodology. This is *not* to suggest, however, that his many experimental findings are all devoid of merit; the individual strength of each purported experimental finding will have to be assessed appropriately. See Schwitzgebel for a sympathetic discussion of selected laboratory training exercises (e.g. Schwitzgebel 2011, Chap. 5). Our criticism will be targeted at the core assumptions and core methodology of Titchenerian psychology.

C. Beenfeldt, *The Philosophical Background and Scientific Legacy of E. B. Titchener's Psychology*, SpringerBriefs in Philosophy, DOI: 10.1007/978-3-319-00242-2_4,

pp. 1, italics removed; also Titchener 1912b, p. 487).[2] In this approach, what at first seems simple is shown "by careful observation to be compound" and it is then "split up into simple parts," and these in turn are split into still simpler parts and so on, "until the science has reached its elements, the simplest things or processes which belong to it, things and processes which cannot be further reduced or more minutely subdivided" (Titchener 1896, pp. 1–2; italics removed). Indeed, according to Titchener, "*all* science begins with analysis" (Titchener 1896, p. 12).

4.2 An Elementary Chemistry of the Mind

The assumed parallel between psychology and the natural sciences is clear. Just as the chemist decomposes an apparent simple substance, such as a ice cube, into the elements of hydrogen and oxygen, and the physicist in turn decomposes these to their subatomic constituents, so too the experimental psychologist seeks to reduce or decompose our complex mental life to its ultimate elementary constituents. In practice, the psychologist "takes a particular consciousness and works over it again and again, phase by phase and process by process, until his analysis can go no further. He is left with certain mental processes which resist analysis, which are absolutely simple in nature, which cannot be reduced, even in part, to other processes" (Titchener 1926, pp. 37–8).

The objective of a given science can thus be generalized as follows: (1) finding that the apparent simple, such as a cube of ice for the chemist or the feeling of anger for the psychologist, is really a complex compound, (2) identifying and isolating its simplest, indivisible constituents, and (3) organizing these constituents systematically, in a manner comparable to what Dmitri Mendeleev did with the chemical elements in what became the modern periodic table.[3] A similar table of the psychic elements would thus be the crowning achievement of a mature science of psychology.

As Titchener asserts, the "first object of the psychologist" is to "ascertain the nature and number of the mental elements" (Titchener 1896, p. 13) and so the psychologist "takes up mental experiences, bit by bit, dividing and subdividing, until the division can go no further. When that point is reached, he has found a conscious element" (Titchener 1896, p. 13).

The notion of psychological elements in Titchenerian psychology is robust, and it is analogous to the notion of elements we find in early scientific chemistry. In his *Outline of Psychology* Titchener offers a list of the known "full resources of the normal mind," which he believes to have shown to amount to at least 42,415 conscious elements (Titchener 1896, p. 67). This includes 30,850 eye sensations,

[2] Alternatively put, "the task of science is to describe; if you are to describe you must analyze" (Titchener 1920a, p. 17) and "[i]t is universally agreed, then, that the first problem of science is analysis" (Titchener 1929, p. 58).

[3] Indeed, "[t]he psychologist arranges the mental elements precisely as the chemist classifies his elementary substances" (Titchener 1926, p. 49).

11,550 ear sensations, four tongue sensations, three skin sensations and a small list of muscle, tendon, joint, alimentary canal, blood vessel, lung, sex organ and other sensations. "Each one of these forty thousand qualities is a conscious element, distinct from all the rest, and altogether simple and unanalyzable" (Titchener 1896, p. 67; italics removed). In the average individual, who often suffers from small abnormalities in their sensory capacities, such as tone deafness, slight colorblindness and so on, the number of elements is smaller but still runs into the tens of thousands (Titchener 1896, pp. 67–8), a figure more than two orders of magnitude greater than the number of currently known chemical elements.[4] Each one of those mental elements can be further "blended or connected with others in various ways, to form perceptions and ideas" (Titchener 1896, p. 67; italics removed).

Modern chemistry provides us with an understanding of the highly complex macroscopic world we inhabit as ultimately decomposable into some finite number of basic elements plus a wide range of potential permutations and combinations.[5] Similarly in Titchnerian psychology, every idea is "built up from impressions" (Titchener 1896, p. 16) and, correspondingly, "every idea can be resolved into elements" (Titchener 1896, p. 26). An idea, then, is "a compound" that "consists of a number of elemental processes, travelling side by side in consciousness" and thus "resembles the compound bodies analyzed in the chemical laboratory" (Titchener 1896, pp. 27–8).

Speaking in a different context, one of Titchener's doctoral students, Grace Adams, captured the reductionistic vision of the project when she observed that "Titchener had foreseen the day when all the data of psychology could be expressed in terms as mathematical as those of a physics equation" (Adams 1930, p. 211).

Unlike chemistry or atomic physics, however, the reduction in psychology is theoretical in a very strong sense, since Titchener denies the possibility (as far as normal human minds are concerned at least) of isolating singular elements experimentally. The reason is that the mental elements cannot ever be quarantined; they cannot ever be experienced in isolation, without the interference of other mental content (Titchener 1914, p. 21).

In his *Lectures on the Experimental Psychology of the Thought-Processes*, Titchener claims that he is ready to "plead guilty to a 'sensationalistic' bias" (Titchener 1909, p. 37). He accepts, as a definition of sensationism,

> the theory that all knowledge originates in sensations; that all cognitions, even reflective ideas and so-called intuitions, can be traced back to elementary sensations (Titchener 1909, p. 23).

For Titchener, the elements of psychology are sensations.

> The idea is a compound; it consists of a number of elemental processes, travelling side by side in consciousness: it therefore resembles the compound bodies analyzed in

[4] New elements are still being discovered with some regularity. Only a few years ago, for example, the discovery of the predicted "superheavy" element 117 was announced.

[5] For the purposes of the present argument, no position is taken on whether this is in fact a correct view of modern scientific chemistry.

> the chemical laboratory. But the sensation resists analysis, just as do the chemical elements oxygen and hydrogen. It stands to the idea as oxygen and hydrogen stand to water. Whatever test we put it to,—however persistent our attempt at analysis and however refined our method of investigation,—we end where we began: the sensation remains precisely what it was before we attacked it (Titchener 1896, pp. 27–8).

"We set out," as Titchener maintained, "from a point of universal agreement. Everyone admits that *sensations* are elementary mental processes" (Titchener 1898, p. 457). What is the nature of these sensations? Characteristically they are individualized sense perception qualities, such as "cold", "blue", "salt" and so on (Titchener 1896, p. 28, 1914, p. 21). Physical atoms combine into molecules and, in a similar manner, atomic sensory qualities combine into real, experienced mental processes. These molecules, in turn, unite to form consciousnesses and a "series of consciousnesses" (Titchener 1914, p. 21) makes up a child or adult or senile mind. These three parts of a whole mental life, when taken together, make up the whole mind of the individual man (Titchener 1914, pp. 21–2). Apart from their lawful interactions, there may be little to say about the essential nature of sensation. As Titchener explained late in life, you "cannot say much about a thing that you regard as ultimate to your science" (Titchener 1917, p. 55).[6]

As we saw earlier, the associationist theory of universal attraction laws (working, ultimately, on the mental elements and compounds) is a derivative notion that requires, and naturally arises from, the more basic assumption of elementism, decompositional reductionism, and sensationism. From this perspective, the laws of association provide the natural answer to the question of how the elements unite into complex structure.

The associationists all agreed on the more fundamental points regarding the laws of association (e.g. that the causal mechanism at work is both lawful and universal), even while they disagreed on all the particulars. Titchener similarly rejected previous formulations of the laws,[7] but he still agreed with the earlier thinkers on the basic issue. Recognizing that in psychology we associate elementary processes (not ideas) and that the association is a psychologically sophisticated connection process (not a "mere juxtaposition") (Titchener 1901–1905, p. 201), it is "one of the fundamental laws of our mental life that all the connections set up between sensations, by their welding together into perceptions and

[6] We are setting aside, here, the relatively late development in Titchener's thought regarding the nature of sensation. Briefly put, Titchener felt impelled by Külpe's definition of a sensation as the sum of its attributes, to accept *sensation attributes* as more elemental that the sensation itself (Boring 1937, p. 476). At an even later point (in fact, starting in 1922 with personal correspondence to Boring) he suggested—as the psychological analogue to his conviction that all of physics was expressible in terms of the three dimensions of space, mass and time—that all sensory attributes were expressible as the *dimensions* of quality, intensity, extensity, protensity (duration), and attensity (clearness, vividness) (Boring 1937, pp. 472–4). We are setting this very late view aside in our discussion of Titchener.

[7] As he put it, "experiments already made furnish additional proof that the old 'laws' of association are psychologically valueless" (Titchener 1920, p. 168).

ideas, tend to persist, even when the original conditions of connection are no longer fulfilled" (Titchener 1901–1905, pp. 200–1). Indeed, in the words of Titchener, we term this law "in conformity with historical usage, the law of the association of ideas" (Titchener 1901–1905, p. 201). The modulating conditions for whether a given association may be established, he further notes, include the frequency, recentness, vividness, and relative position in a series, of a given connection (Titchener 1901–1905, p. 201).[8]

In general, the aim of scientific psychology is to answer three questions about its subject matter: the what, the how, and the why. The what question is answered by analyzing a mental experience into its elements, the how question is answered by *formulating the laws of connection of these elements*, and the why question is answered by explaining mental processes in terms of their parallel processes[9] in the nervous system (Titchener 1926, p. 41, italics added). For both Titchenerian psychology and British associationism, the laws of association furnish an explanation of how the mental mechanics operate.

4.3 Dynamic Mental "Atoms"

Consider, now, Titchener's definition of the conscious or mental elements as "those mental processes which cannot be further analyzed, which are absolutely simple in nature, and which consequently cannot be reduced, even in part, to other processes" (Titchener 1896, p. 13). The operative word here is *processes* (Titchener 1899, pp. 293–6). In psychology, as he explained, "we observe events, occurrences, happenings, goings on, processes: never things" (Titchener 1914, p. 7).This is puzzling. After all, one would expect that given Titchener's reduction of mental life to elemental constituents—in a carefully laid out project that paralleled the recognized work in scientific chemistry of identifying, systematizing and cataloging the elements of matter—he is really assuming that some kind of mental atoms serve as the comparable constituents in psychology. In the final analysis, this is indeed what he is doing. To see the problem, however, consider his criticism of outright mental atomism. In brief, he thinks that it ignores what he regards as the "most striking fact about the world of human experience," namely the fact of *change*: "[n]othing," he observes, "stands still" here (Titchener 1926, p. 15). Indeed, he claims, the term "process" itself was "imported into modern psychology by way of reaction against the preceding psychological atomism" (Titchener 1899, p. 295).

Yet, what exactly does this process perspective on the elements and compounds of mental life amount to? Not much more, it seems, than the insistence that our mental life is dynamic rather than static and unchanging. As Titchener sees it, a thing is permanent, relatively unchanging and definitely marked off from other

[8] The quotes are from the *Student's Manual*, Volume I Qualitative Experiments: Part I.

[9] The idea, here, seems to be that of psycho-physical parallelism.

things, whereas a process is a "moving forward", a continuous operation, a constant process of change (Titchener 1896, p. 5). Unlike a child's puzzle-map, the elemental processes do not fit neatly together "side to side and angle to angle," they rather "flow together, mix together, overlapping, reinforcing, modifying or arresting one another, in obedience to certain psychological laws" (Titchener 1896, p. 15).

Take the idea of a tree. Unreflectively, we may think that this mental item is a stable thing. Upon close examination, however, it is revealed as a complex mixture "containing a number of colors, a number of lights and shades, a number of forms" (Titchener 1896, p. 6). When we pay attention to the idea, we give varying attention and emphasis to these different constituents of it—"[n]ow the form of the tree is uppermost in our mind, now its shadow, now the stickiness of its buds, now some incident connected with it" and so on (Titchener 1896, p. 7). In short, the idea changes and is therefore not a thing: "it does not stand, like the rock" solid and unchanging (Titchener 1896, p. 7). It is a dynamic process, like the tumultuous waves crashing upon the rock, or like heat, or decomposition, or like the erosion of a cliff.

Observe, however, that Titchener is defining his own dynamic account of the elements as a contrast to a stability account that must be recognized as a straw man opponent. Consider the following analogy: A large wooden bench in my garden is "permanent, relatively unchanging and definitely marked"[10] if our standard of assessment is unaided human perception and the time frames of everyday human experience. Even after several years have passed, let us imagine, only a few scratches have appeared on its surface. An apple in the same garden, however, is clearly not "permanent [and] relatively unchanging" by the same standard of macroscopic change. If we leave it outside for a few days or weeks, its appearance and structure will have been altered markedly. Our assessment of the wooden bench as completely inactive, however, depends crucially on the grossness of grain of the perspective we have adopted. On a sub-macroscopic level, the bench is a bustling intersection of active, and causally interacting, atomic constituents.

Similarly if we imagine speeding up time sufficiently, we find this apparently permanent object undergoing change and alteration. Whatever the ultimate material constituents of the universe may be, quarks, sub-quarks or something else, no physicist assumes that they are inactive and unchanging. It is therefore not very convincing to maintain, as Titchener does, that the psychologist eschews mental atomics because he, unlike the chemist with his hydrogen and oxygen, recognizes that *his* elements are dynamic rather than immovable and unchanging.

Titchener's argument also fails in another respect. The problem is that, according to the theory, nobody has ever experienced an individual, elemental sensation: nothing that we experience is atomic. "[N]o concrete mental process, no idea or feeling that we actually experience as part of consciousness, is a simple process"

[10] Setting aside deliberate human action, environmental force majeure, and so on.

because all alike are "made up of a number of really simple processes blended together" (Titchener 1914, p. 21). If we grant that nobody has ever *actually* experienced an elemental sensation, what we do experience (i.e. mental *compounds*) cannot simply be taken to establish the nature of those elements. In other words, we cannot simply assume that the nature of the elements is precisely the same as that of the compounds we do experience: the properties of a compound (such as those of the idea of a tree), after all, may differ substantially from the properties of its building-block constituents—just as water (the compound) differs from the isolated elements of hydrogen and oxygen (the simples).[11]

Notice, finally, that the items actually picked by Titchener as parallels to the fundamental process characters of the mental elements undermine his process claim. The mental processes, we are told, are like waves upon rock, heat, decomposition, or the erosion of a cliff. However, waves consist of atoms, rocks consist of atoms, and the phenomenon of waves-hitting-upon-rock can be understood in terms of atomic-level interactions and so, trivially, can the crumbling of a cliff. Temperature (as well as heat) is understood in terms of the properties of atoms such as kinetic energy and the flow of energy between systems. Decomposition is not understood as an elemental process either. The concept "decomposition" is an action concept that denotes a special type of action, process, or change that a physical object undergoes. As a forensic scientist might explain, a body undergoing decomposition is a physical object that is in the process of breaking down on the levels of tissue, cell, biopolymer, and so on (typically while also being consumed by insects and bacteria). Without the relevant organic component that actually disintegrate, there could be no such free-flowing decomposition taking place.

In all of these cases, the atomic or cellular view is avoided by virtue of an incomplete analysis. They are avoided, in other words, by treating waves-hitting-rock, heat, and decomposition as incompletely analyzed process-elements, when they are in fact easily shown to be mere activities, interactions, properties or states, of underlying cellular, or molecular, or atomic structures. Thus, Titchener's insistence on upholding mental processes, as opposed to mental atoms, rings hollow indeed.

Furthermore, even as he maintains that those elements are process-like rather than atomic in nature, he consistently treats them as if they are much more tangible than a genuine process account could ever countenance. Consider, for example, his illustration of the mind in *Outline of Psychology* (Titchener 1896, p. 10), in which mental processes are depicted (for the pedagogical benefit of the beginning student of psychology) as individual strings together making up the total mental life of an individual like so many fibrous strands of a rope. This is an elementist image of the mind, and a palpably corporeal one at that. Perhaps, then, the process notion was maintained as a hand-waving gesture to deflect criticism of mental atomism (Titchener 1899, pp. 293–6). Perhaps it was chosen because it

[11] For Titchener the "elements are posited by theory" (Hatfield 2005, p. 267).

seemed to conform better to the actual experiences of mental life (although, as we have seen, this is strictly speaking irrelevant as those experiences are always compounds, never the isolated elements themselves) or perhaps it was preferred out of an incomplete late nineteenth-century understanding of the microphysics on the hard science side of the parallel between psychology and the natural sciences. Perhaps there were other reasons[12] and perhaps it was a combination of the above.

4.4 The Elementary Units as Sensationistic

Here I will comment, finally, one the classes or types of elementary units recognized by Titchenerian psychology. In his 1909 work, *A Text-Book of Psychology*, Titchener recognizes three such classes: sensations, images and affections.[13] Sensations are the characteristic elements of perceptions, the "sights and sounds" we enjoy in response to environmental stimulation (Titchener 1909, p. 48). Images are the constituents of our ideas and they are so similar to sensations that the two "are not seldom confused" (Titchener 1909, p. 48). Affections are the characteristic elements of emotions such as love, hate, joy and sorrow (Titchener 1909, p. 48).

Images, Titchener suggests in the *Text-Book*, may simply be sensations. Usually a sensation and its corresponding image are said to differ in that the image is more pale, less intense and of a shorter duration (Titchener 1926, p. 198), but these are all "differences of degree, and not of kind" (Titchener 1926, p. 198). This account suggests the possibility of combining images and sensations into one elemental group, presumably with images (the qualitatively inferior copy of sensations) understood as the dependent product of sensations. Titchener considers a range of experimental findings (Titchener 1896, pp. 29–30) that lend weight to this hypothesis. Yet, he reaches no firm conclusion on the issue (Titchener 1926, pp. 197–200). Six years later, in *A Beginner's Psychology*, he writes that it is "very doubtful if there is any real psychological difference between sensation and image" (Titchener 1920a, p. 73, italics removed).

Whatever the ultimate number of elemental categories turns out to be, it is clear that in Titchener's view, sensations are the ruling constituents in human mental life. They are ultimately responsible for perceptions (a "group of the sensations") (Titchener 1901–1905, p. 127), images (which differ from sensations in *degree*, not kind),[14] and ideas (which differ "from a perception only by the fact that [they

[12] For instance, Wundt's influence (Greenwood 2003). Alternatively, Boring (1937, pp. 472–4) suggests that Titchener's stipulation of sensations as the elementary units of psychology was the reason for this.

[13] We will discuss affections late. Here, we simply note that Titchener favors the classification of them as a separate conscious element "ranged alongside of sensation in the composition of consciousness" (Titchener 1901–1905, p. 91).

[14] Titchener offered some finer distinctions with respect to sensations and images in a note (Titchener 1904) published a few years later.

are] made up wholly of images") (Titchener 1926, p. 376). Indeed, Titchener exhorts us to recognize that there "is no fundamental psychological difference between the perception and the idea" (Titchener 1896, p. 148).[15] We might, of course, be tempted by common usage "to employ 'perception' to denote what is now before us, and 'idea' to denote what is remembered or imagined" (Titchener 1896, pp. 148–9) but we should resist this temptation and "remind ourselves that, in principle, the two processes are one and the same" (Titchener 1896, p. 149). Indeed, Titchener goes as far as to adopt the policy of using the words *indiscriminately* (Titchener 1896, p. 149).[16] Clearly, Titchenerian psychology is deeply sensationistic in nature.

4.5 Elementism, Reductionism, Sensationism, and Association

Let us now draw together the evidence offered in the previous section in support of the claim that introspectionism should be understood as an associationist project through and through. The common aim and agenda of both associationism and introspectionism was to seek (1) a reductive decomposition of human mental life into (2) elements that ultimately are (3) sensationistic in nature and (4) follow certain law-like causal regularities (e.g. the classical laws of association).

1. The structural psychologist must, as Titchener stated,

   > be able to say: "Give me my elements, and let me bring them together under the psychophysical conditions of mentality at large, and I will guarantee to show you the adult mind, as a structure, with no omissions and no superfluity (Titchener 1899, p. 294, italics removed).

Going the other way, the task of psychology is to take up "mental experiences, bit by bit, dividing and subdividing, until the division can go no further. When that point is reached … a conscious element" has been found (Titchener 1896, p. 13).

2. The "first object of the psychologist" is to "ascertain the nature and number of the mental elements" (Titchener 1896, p. 13, italics removed).
3. The sensation is "the structural unit or structural element" of consciousness and so "if we wish to understand the make-up of mind we must know all about these sensations" (Titchener 1901–1905, pp. 1–2).

---

[15] Remembering that perception, in turn, is "primarily a group of sensations" (Titchener 1901–1905, p. 127).

[16] As an illustration of this approach, consider Titchener's analysis of the abstract idea of "hour." Examining his own idea of "hour," he explains that this mental item simply "*consists of* the picture of a small outline square drawn on a white background; and this square is one of the squares of the daily report-cards upon which the marks for every hour's work were entered at the first school that he attended" (Titchener 1914, p. 221, italics added).

4. It is "one of the fundamental laws of our mental life that all the connections set up between sensations, by their welding together into perceptions and ideas, tend to persist" (Titchener 1901–1905, p. 200).

Like John Stuart Mill, Titchener regarded psychology to be a sort of mental chemistry.

> The psychologist arranges the mental elements precisely as the chemist classifies his elementary substances. The chemical elements are divided, for instance, into metals and non-metals. The metals have a high power of reflecting light; they are opaque; they are good conductors of heat and electricity; they have high specific gravities. So they are set off, as a group, from the non-metals. These latter, again, include both gaseous and solid elements. That is to say, the chemical elements possess certain properties or attributes, by means of which they may be distinguished and arranged. It is just the same with the mental elements (Titchener 1926, pp. 49–50).

We might say, then, that introspectionism was laboratory-based form of British associationism.

4.6 An Englishman Representing the British Tradition

The assessment of introspectionism as a laboratory-based form of British associationism, however, diverges sharply from Boring's characterization of Titchener as "an Englishman who represented the *German* psychological tradition in America" (Boring 1950, p. 410, italics added). However, it is likely that Boring's estimation of Titchener's intellectual home soil was simply a product of what we have called the traditional account. Rather than representing the German tradition, Titchener in fact had a fundamental theoretical commitment to the central tenets of *British* associationism. He might have picked up the genuinely empiricist ideas in Wundt's work (e.g. Wundt's apparent elementism in psychophysics), recast other ideas in the empiricist mold (e.g. the highly circumscribed introspective analysis which, in Titchener's hands, was transformed into the universal methodology of psychology), and discarded or downplayed the rest (e.g. Wundt's *Völkerpsychologie*).[17] Perhaps he was genuinely convinced that he had distilled a purified version of Wundtian psychology—purified, that is, from the Wundtian ideas that were not British associationism.[18]

---

[17] In 1921 Titchener published a biographical essay about Wundt (who had died the previous year). Here, he attempts to answer the question: whence "did Wundt derive his idea of an experimental psychology?" (Titchener 1921, p. 164). His answer is that the "proximate source of Wundt's idea is patent"—it is, namely, the sixth book of John Stuart Mill's *Logic*, in which the associationist doctrine of mental chemistry is expounded. The cardinal difference between Mill and Wundt is that "Mill talked about experiments and Wundt carried them out" (Titchener 1921, p. 165).

[18] Leahey similarly argues for a deliberate filtration of Wundt on the part of Titchener, though he emphasizes Titchener's positivist philosophy of science more than his associationist psychology (Leahey 1981).

While we declined to take a stand on the dispute regarding the traditional/modern account of Wundt, the present discussion should be of use to advocates of the modern account, in that it answers the following question: If Wundt was not the near-associationist that Boring claims, how then did Titchener come to be? The present claim is that there is no need for the traditional Wundt-as-proto-Titchener account to answer this question. If one accepts the analysis of introspectionism as simply a developed form of associationism—and then adds the biographical supposition that Titchener was already thoroughly familiar with associationist thought before he began developing his distinctive psychological approach under Wundt's tutelage—one can explain the development of Titchener's approach to psychology without having to assume Boring's account of Wundt.

The aforementioned biographical supposition can be supported by evidence quite easily.[19] Titchener went to Oxford in 1885 and spent his first four years there studying philosophy.[20] In his study of this subject included readings of James Mill, John Stuart Mill, and Alexander Bain. Indeed, it was "a paragraph in James Mill" that set him, as he autobiographically put it, "on the introspective track" (Titchener 1920a, p. vii)—a track that, we have seen, stretched all the way through his career in experimental psychology.

Other influences include Thomas Hobbes ("perhaps the greatest of English philosophers") (Titchener 1914, p. 135), as well as David Hume ("the great philosopher"), whose reduction of the human mind to the distinct elements of impression and ideas is mentioned approvingly by Titchener (1920a, p. 73). He makes clear in the Preface to *Outline of Psychology*, that the "general standpoint of the book is that of traditional English psychology" (Titchener 1896, p. vi). Pillsbury notes that "in the earlier years" Titchener, spoke frequently of Mill's mental chemistry and advocated a search for means to analyzing the given concrete process into elements (Pillsbury 1928, p. 97).[21] In fact, Titchener's "belief in introspection," can be dated rather accurately, "for it was in 1888," as he later recalled, "when for the first time I was reading James Mill's *Analysis*, that the conviction flashed upon me—'you can test all this for yourself!'—and I have never lost it since" (Titchener 1909, p. 96).[22]

[19] Danziger, it should be noted, called attention to this biographical fact some thirty years ago (Danziger 1980, p. 246).

[20] Titchener was, at Brasenose College, a Senior Scholar in Philosophy, as well as the Classics and Senior Hulmian Exhibitioner (Pillsbury 1928, p. 95).

[21] Pillsbury notes that Titchener finally seems to have abandoned his allegiance to mental chemistry very late in life (Pillsbury 1928, p. 97). As far as Titchener's very late views are concerned, we take no position on this issue.

[22] Titchener came, as a student at Oxford, "under the influence of the English psychology, apparently not so very much modified from the time of the Mills" (Pillsbury 1928, p. 96).

References

Adams G (1930) Psychology: science or superstition? Covici-Friede, New York
Boring EG (1927) Edward Bradford Titchener. Am J Psychol 38(4):1867–1927
Boring EG (1937) Titchener and the existential. Am J Psychol 50(1/4):473–483
Greenwood JD (2003) Wundt, Völkerpsychologie, and experimentalsocial psychology. Hist Psychol6(1):70–88
Boring EG (1950) A history of experimental psychology. Appleton-Century-Crofts, New York
Danziger K (1980) The history of introspection reconsidered. J Hist Behav Sci 16(3):241–262
Evans RB (1984) Titchener and American experimental psychology. Rev Hist Psicol 5(1–2):117–125
Hatfield G (2005) Introspective evidence in psychology. In: Achinstein (ed) Scientific evidence: philosophical theories and applications. Johns Hopkins University Press, Baltimore
Leahey TH (1981) The mistaken mirror: on Wundt's and Titchener's psychologies. J Hist Behav Sci 17(2):273–282
Pillsbury WB (1928) The psychology of Edward Bradford Titchener. Philos Rev 37(2):95–108
Schwitzgebel E (2011) Perplexities of consciousness. MIT Press, Cambridge
Titchener EB (1896) Outline of psychology. The Macmillan Company, New York
Titchener EB (1898) The postulates of a structural psychology. Philos Rev 7(5):449–465
Titchener EB (1899) Structural and functional psychology. Philos Rev 8(3):290–299
Titchener EB (1901–1905) Experimental psychology: a manual of laboratory practice. The Macmillan Company, New York
Titchener EB (1904) Organic images. J Philos Psychol Sci Methods 1(2):36–40
Titchener EB (1909) Lecture on the experimental psychology of the thought-processes. The Macmillan Company, New York
Titchener EB (1912) The schema of introspection. Am J Psychol 23(4):485–508
Titchener EB (1914) A primer of psychology. The Macmillan Company, New York
Titchener EB (1917) The psychological concept of clearness. Psychol Rev 24(1):43–61
Titchener EB (1920) A beginner's psychology. The Macmillan Company, New York
Titchener EB (1921) Wilhelm Wundt. Am J Psychol 32(2):161–178
Titchener EB (1926) A text-book of psychology. The Macmillan Company, New York
Titchener EB (1929) Systematic psychology prolegomena. Cornell University Press, New York

Part III
The Preeminence of Analysis, Not Introspection

Chapter 5
The Decline and Fall of Introspectionism

5.1 No Mere Issue of Reliability

Why did introspectionism fall? "The approach failed," a typical philosophy of science explanation will inform us, "because its methodology lacked reliability." In *Pathways to Knowledge*, for example, Goldman offers the following account of the failure of what he calls classical introspectionism:

> Subjects trained in different laboratories delivered different judgments about their consciousness. This is not technically a contradiction, since each was talking about his or her *own* consciousness. But if we add the (implicit) premise that everyone's consciousness has the same general features, inconsistency follows. This was why introspection was abandoned at that historical juncture; it was a problem, at least an apparent problem, of reliability (Goldman 2004, p. 112).

For Goldman, the defects of introspectionism could have been corrected, it seems, by limiting the scope of the questions asked during an introspective investigation. "When," as he puts it, "introspection is confined to a more modest range of questions, as it is today, it has not been shown to be unreliable in such a domain" (Goldman 2004, p. 112). In other words, introspectionism failed because it carried out an investigative approach that presented experimental subjects with questions having *too wide* a scope to ensure scientific reliability. However, this basic approach can be rendered viable by the introduction of an adjunctive qualification that constricts the scope of the questions posed.

Although Goldman does not specify how his scope constriction would apply to any historical introspectionist experiments, this posited corrective alone cannot salvage the system. The suggestion, in fact, reflects a considerable underestimation of the pernicious methodological problems that introspectionism was beset by. It also underestimates the extent to which introspectionism *did* aim at confining subject queries to a very "modest range of questions" and *did* employ a narrow, circumspect, almost surgical, terminology in an attempt to capture the minute details

C. Beenfeldt, *The Philosophical Background and Scientific Legacy of E. B. Titchener's Psychology*, SpringerBriefs in Philosophy, DOI: 10.1007/978-3-319-00242-2_5,

of a given conscious process. This was a central feature of Titchener's method and it was an important part of the problem with introspectionism.

It is not just among philosophers of science, however, that we find an inadequate understanding of the problems with Titchenerian methodology. It is also the accepted view voiced in history of psychology textbooks. As Costall describes, the

> textbook histories usually just mention the fact that introspection yielded inconsistent results, and then conclude, on this basis alone, that the method was clearly invalid. Yet, even by the standards of fictionalized, disciplinary history, this is not good enough (Costall 2006, p. 647).

Why is this not good enough as an explanation of the apparent failure of introspection in psychology?

> Failures of agreement among researchers are hardly unique to introspection or to the discipline of psychology, and do not necessarily imply a lack of scientific rigor or honesty, or even an eventual "dead-end" (Costall 2006, p. 647).

This is coherent with Danziger's observation that situations in which "the results from one laboratory are at variance with those from other laboratories are far from unknown in the history of science" (Danziger 2001, p. 7890).

Returning to the attempt to explain the failures of introspectionism, Eric Schwitzgebel offers an account according to which there "were many reasons for the death of classical introspective training" (Schwitzgebel 2005, p. 8). The four reasons he lists are: (1) "legitimate concerns and objections" raised by theoretically opposed psychologists that were, however, "somewhat overplayed;" (2) the fact that the program "yielded few socially valuable results and bogged down in sterile debates;" (3) the "exciting and seemingly much more useful research prospects" of behaviorism and functional psychology; and (4) the "tiresome" nature of introspective training (Schwitzgebel 2005, "Concluding Discussion").

There is some truth in each of these four points, although there is also a deeper problem with Schwitzgebel's analysis. According to Schwitzgebel, Titchenerian psychology is a prime representative of genuinely introspection-based psychology, "the *archetypal* method of introspective psychology" (Schwitzgebel 2007c, p. 225). Furthermore, because he does not want to simply reject introspection, Schwitzgebel apprehensively defends Titchenerian introspective training and "recommend[s] that we consider introspective training as a potential response to [the] difficulty" created by the tension between the necessity of relying on introspection in the study of consciousness on the one hand and the potential of unreliable introspective reports on the other hand (Schwitzgebel 2004, p. 58). Schwitzgebel thus endorses, in certain respects, a position that he traces back to introspectionism (Schwitzgebel 2002, p. 37). As we shall see, however, Titchenerian psychology is not a good representative (let alone archetype) of a genuinely introspection-based approach in psychology, and Titchenerian training was a major part of the *problem*, not something we should reintroduce in consciousness studies today.

5.2 Special Training Required

Returning to the false textbook wisdom mention by Costall, let us begin to correct it by observing that "experimental introspection" or "systematic introspection," or simply "introspection," did not in fact *mean* introspection *simpliciter* for Titchener. It meant something different. In Titchener's view, the term "introspection," if it is used validly in a scientific context, can only mean a certain systematic and (as we shall see) heavily theory-laden form of psychological enquiry. As I will argue, the proper term for this approach is not "introspection" in an unqualified sense, but rather "psychological analysis," "analytic attention," or simply "analysis."

The employment of correct introspective analysis or analytical attention, according to Titchener, requires the full attention, the exercise of the highest possible degree of concentration, on the part of the subject (Titchener 1926, p. 24). The motivation for this requirement was that introspective attention to the conscious process was held to be inherently retrospective. As Titchener wrote, scientific introspection is a "post mortem examination" (Titchener 1914, p. 28; italics removed), in the sense that the subject must always recall the experience that he has just undergone. If sufficient concentration is exercised, however, the experimenter will, by the vicarious means of the participant's inner eye, be able to obtain a "photographically accurate" record of the experienced mental process (Titchener 1926, p. 24).[1]

Exacting levels of attention and concentration and the ability to produce photographically accurate records of the minutia of one's experience, however, lie far outside the capabilities of the random and untrained subject. This requires *trained experts*, i.e. subjects who have habituated the performance of such observations to the point where execution of the task itself does not corrupt the gathered material. Introspection, if it is to have any standing at all, must therefore be the product of "the ingrained habit of observation that has been molded in the laboratory" (Titchener 1912a, p. 444). "You yourself know very well," as Titchener wrote in personal correspondence with Boring, "how absurdly unreliable are the judgments of perfectly competent persons who have had general but not special training" (quoted in Boring 1937, p. 475).

What sort of special training did Titchener have in mind? For this, we turn to the famous "Titchener Manuals." These works were "standard manuals for laboratory practice and were widely used long after Titchener's death in 1927."[2] The

[1] As Schwitzgebel explains in his examination of Titchener's *Manual*, the difficulties in achieving the desired effect "include maintaining consistent attention, avoiding bias, knowing what to look for, and parsing the complexity of experience as it flows rapidly past" (Schwitzgebel 2004, p. 61).

[2] Evans, "The Scientific and Psychological Positions of E. B. Titchener" (Evans 1990, p. 7). On the same page, Evans notes that "[e]ven in departments where the content of Titchener's psychology was rejected, his *Experimental Psychology* was the standard text in the experimental course."

manuals, formally entitled *Experimental Psychology*, were issued in four volumes between 1901 and 1905[3] and were works upon which "[g]enerations of psychologists cut their experimental teeth."[4] They were studied and the experiments described in them were carefully worked through by countless psychology students around the turn of the century. Even psychologists, such as Watson, who were to become famous opponents of Titchener, were trained in experimental technique with this text in hand.[5] The purpose of the works was to inculcate in the student of psychology the established laboratory practice of the field.

As we open page one of the work,[6] we read that a psychological experiment is "a dissection of consciousness, an analysis of a piece of the mental mechanism" (Titchener 1901–1905, p. xiii).[7] Moving from the Introduction to the first paragraph of Chap. 1, we read that the aim of this analysis is "to split up a complex whole into simpler parts, and so ultimately to reduce it to its elements, the very simplest factors of which it is compounded." The next line reads: "You can see, then, even before you have performed a psychological experiment, that an analysis of some particular section of consciousness means the splitting-up of it into simpler and simpler mental processes, until finally it reveals itself as a complex of elementary processes, beyond which we cannot go" (Titchener 1901–1905, p. 1). Let me repeat that: "You can see, then, *even before you have performed a psychological experiment*, that …" On the very first page[8] of Titchener's famous laboratory manual, the student is instructed that a psychological experiment is nothing but … what? It is nothing but an act of *psychological analysis*, a process of seeking sensory elements via decompositional reductionism.[9] This, then, is how the ingrained habit of observation was to be created.

---

[3] The four volumes are: *Volume I Qualitative Experiments: Part I. Student's Manual* (New York: Macmillan, 1901); *Volume I Qualitative Experiments: Part II. Instructor's Manual* (New York: Macmillan, 1901); *Volume II Quantitative Experiments, Part II. Student's Manual* (New York: Macmillan, 1905); and *Volume II Quantitative Experiments: Part II. Instructor's Manual* (New York: Macmillan, 1905). We here adopt the convention of citing these closely connected works as Titchener 1901–1905. Quotes in this book are from the *Student's Manual*, Volume I Qualitative Experiments: Part I.

[4] Evans, "The Scientific and Psychological Positions of E. B. Titchener" (Evans 1990, p. 7).

[5] Berman and Lyons (2007, pp. 5-15) provide some details about Watson's early work as an introspectionist experimenter, as well as his growing misgiving about this work. They also discuss the warm and long-lasting Watson-Titchener friendship.

[6] *Volume I Qualitative Experiments: Part I. Student's Manual* (New York: Macmillan, 1901).

[7] This is from the section entitled "Introduction: Directions to students".

[8] To be precise, the very first page of the main text in the *Student's Manual*.

[9] In the next paragraph sensationism is introduced. "The sensation, then, is the structural unit or structural element of these consciousnesses … just as the cell … is the structural element of our bodily tissues" (Titchener 1901–1905, pp. 1-2).

On page two, Titchener explains why it would be a foolish waste of time for the student of psychology to try to reach, for himself, the conclusion that mental phenomena invariably split into elemental sensations:

> We might ... take some complex consciousness, such as the memory-consciousness, and analyze it for ourselves, in order to test the statement that it can be split up into sensation elements. But this would really be wasting time. For we do not yet know what a sensation is; we should not be able to recognize it, if we came across it. We should be struggling, ignorantly, to solve a difficult problem by 'common sense,' and neglecting the skilled work of those who have attacked the problem before us,—work that would spare us much futile effort and many errors. Or, to put it differently, instead of getting a map and following the high-road, we should be pushing across country, by the help of a pocket compass, towards a town which we know to lie in a certain direction. What we shall rather do, then, is to begin with an examination of sensations, in light of the experimental work already done. When we are thoroughly familiar with the mental elements, we shall know what sort of thing to look for in our analyses of complex processes: our dissection will be surer and safer, and we shall be better able to grasp the pattern of these complexes, to trace out the peculiar forms of mental connection (Titchener 1901–1905, p. 2).

With apologies to Lord Tennyson, one might here say that "[s]ome one had blunder'd" and, continuing those famous lines, that "[t]heirs not to make reply, [t]heirs not to reason why"—theirs but to do the experiments and find the mental elements.[10]

5.3 The Experimenter as a Scientific Apparatus

The motivation for this methodological indoctrination was not senseless dogmatism on Titchener's part.[11] Rather, the reason was that experimental introspection in the Titchenerian laboratory was treated as a kind of *scientific apparatus*, a device expressly designed to perform a specific task, to produce a specific type of data for which it had been designed and calibrated beforehand. As Titchener puts it, the limits of introspection are like the limits of a microscope or of a camera (Titchener 1912b, p. 498). Elaborating on the point, we might observe that one can obtain only a very specific type of data with any given optical microscope. Put simply, one can detect the visual characteristics, at a certain level of optical

[10] In his attempt to put forward the most plausible and commonsensical account of the introspectionist methodology, English explains in 1921—as *the first rule of psychological experimentation*—that one "describe the constituent features of the experience in terms that resist further analysis," a description "in terms of the part-processes which cannot be thought of as being themselves made up of smaller or simpler part processes" (English 1921, p. 406). Here, again, psychological analysis is regarded as the bedrock of an introspection-oriented approach to psychological research.

[11] Though Titchener seems not to have minded being perceived as dogmatic. According to Boring, Titchener had an "invariable custom" of lecturing in an Oxford master's gown because, he held, it "confers the right to be dogmatic" (Boring 1927, p. 492).

magnification, of suitably lit material samples that fit onto the microscope stage. Trivially, our 100 × magnification value optical microscope will allow us to see neither atoms, nor black holes, nor the ecological group dynamics of African zebra herds. Similarly with introspection, one "can observe only what it observable" (Titchener 1912b, p. 498). The significance of this claim becomes clear when we then ask, *what is taken to be observable?* Here, the basic premises of elementism, reductionism and sensationism decide the issue.

According to Titchenerian psychology, we can observe only sensations and affections, or rather—since these are in fact unobservable in their pure form—compounds of these, and most notably of the former. When we attend to an abstract idea, we invariably find simply images. When we attend to images, we invariably find compounded sensations. When we attend to perception, we invariably find groups of sensations. When we try to isolate those sensations, we get to simpler and simpler sensations, and we ultimately infer some 40,000 elemental items. We may not experience them "pure," but we get as close to this as the magnification range of our introspective microscope permits.

Titchener's putatively introspective methodology—i.e. the process of scientific psychological analysis—was thus a theoretical derivative investigative apparatus: it was a tool designed for a task that was set long beforehand, namely the experimental confirmation of the classical empiricist perspective on the mental.

Recall that for Titchener, scientific psychology must fall into line with the remaining sciences that all "find their source in analysis" (Titchener 1896, p. 1, italics removed). This means that psychology *must* perforce advance by way of a reductive, decompositional search for the elemental, atomic, mental constituents that, by causal compounding, make up the structure of human psychology. To fail to proceed in this way is not to carry out a poor scientific investigation—it is not to carry out a scientific investigation of human psychology *at all*. What is recognized as introspection today, by philosophers, psychologists, and ordinary people alike,[12] was regarded as a wholly illegitimate form of enquiry by Titchener. It is an activity that belongs forever far outside the walls of proper science.

5.4 All Science Begins with Analysis

This all follows from Titchener's general philosophy of science. As he asserted, "all science begins with analysis" (Titchener 1896, p. 12). In scientific psychology this means that one "takes a particular consciousness and works over it again and again, phase by phase and process by process, until [the] analysis can go no further" (Titchener 1926, p. 37). Because the process of analysis is the "sole gateway to psychology" (Titchener 1914, p. 32), "our one reliable method of knowing ourselves" (Titchener 1914, p. 32), and the "final and only court of appeal" in matters of

[12] More on the topic of different notions of introspection in the next chapter.

psychology (Titchener 1896, p. 341), and since this method is essentially tied to elementist, reductionist, and sensationist commitments, psychology itself is interminably chained to those assumptions *regardless* of the incoming experimental evidence.

"If the object of the psychologist is to know mind, to understand mind, then it seems to me," as Titchener wrote in a paper reviewing experimental work produced in psychology from 1900 to 1910, "that the *only course* is to pull mind to pieces, and to scrutinize the fragments as minutely as possible" (Titchener 1910, p. 421, italics added). If science simply *is* the practice of analysis within some suitable domain of investigation, then no scientific evidence can ever disconfirm the practice of analysis. What evidence, after all, could possibly invalidate a "final and only court of appeals?" No testimony can overturn the legitimacy of the very court that presides over its recognition as *forensic evidence*.

Titchener exhorted the new student of psychology not to attempt to reach his destination by pushing across country and getting lost in the wilderness, but to accept a map drawn on the basis of the assumption of sensationism, elementism, and reductionism, and, with that map in hand, to travel on the scientific "high-road" on which he will join the existing community of investigators who search for the mental elements and compounds. As Danziger observed, "[i]t was in Titchener's circle that the emphasis on the trained observer became a matter of principle" (Danziger 1980, p. 246). If we call this the *positive strategy* (in the logical sense) of regimentation into Tichenerian psychology, the *negative strategy* was the famous introspectionist tactic of rejecting discordant observational data as the product of the *stimulus error*.[13, 14]

Imagine for a moment that someone is conducting a laboratory study of human perception where the experimental setup requires the subject to visually discriminate colored pieces of paper. As a subject, you would be required to, very carefully, attend to your experience and to describe this experience to the experiment-leader afterwards. You might, commonsensically, think of saying something like "I just saw a red piece of paper," but that would be all wrong. It would be wrong because you spoke of seeing red *paper*. What you should have said was that you experienced a so-and-so sensation of red (Titchener 1912b, pp. 488–489). In other words, in your erroneous report you clearly attended, *not* to the red sensation, but to the red paper. And that is an error. It is the stimulus error.

The observer in a psychological experiment falls into [the stimulus error], as we all know, when he exchanges the attitude of descriptive psychology for that of common sense or of natural science; in the typical case, when he attends not to 'sensation' but to 'stimulus' (Titchener 1912b, p. 488).

---

[13] See Boring's discussion of the stimulus error in his paper on this topic (Boring 1921). Here, Boring argues that Titchener first uses the term in *Experimental Psychology* (Boring 1921, p. 451 footnote).

[14] Schwitzgebel seems to view Titchener's use of the stimulus error favorably, as a largely sensible caution against lapsing in one's efforts to study sensory experience (Schwitzgebel 2005, "Concluding Discussion"). This favorable estimate is echoed in Schwitzgebel 2007, p. 52 and Schwitzgebel 2011, p. 148. The aim of the present discussion is to show a *darker side* of this Titchenerian strategy.

Setting aside abnormal conditions like neurological pathologies or the presence of psychoactive drugs in the system, one might object that we never perceive the sensations, but always the stimulus: what we see is a world of things, a complex physical environment of red pieces of paper, green lawns, brown wooden desks, and so on. Yet this view, says Titchener, is mere *common sense* and it has no place in proper experimental psychology.

5.5 No Introspection Through the Glass of Meaning

The science of psychology needs the investigative approach of Titchenerian experimental introspection precisely because without it, the observer is "warped and [biased] by common sense" (Titchener 1920a, p. 23). Such warping and biasing may be caused by our knowledge of natural science, by logical reflection, and by considerations of value (Titchener 1912b, p. 488–489). "[I]ntrospection through the glass of meaning," as Titchener put it, "that is the besetting sin of the descriptive psychologist" (Titchener 1899, p. 291). Acts of introspection that fail to comply with this standard are summarily dismissed as "scientifically illegitimate" or "wholly imaginary" procedures (Titchener 1912b, p. 485) and even as "perverted introspection" (Titchener 1899, p. 292).

We are introspecting correctly—we are, as experimental subjects, "playing the game" as Robert Woodworth put it in a critical tone (Woodworth 1915, p. 7)—when we eschew values, interpretation and even meaning, i.e. when we no longer describe seeing a red piece of paper as seeing a red piece of paper, but instead produce laboratory data that comply with the adopted assignment of breaking consciousness down into simpler processes and ultimately into elemental sensations. From our knowledge of the ontogenesis of more mature sciences like Newtonian physics or Mendeleevian chemistry, we can feel assured that this is the yellow brick road that psychology also must follow. "All science" after all "begins with analysis."[15] And so, of course, must mental chemistry.

Consider, as an example, Titchener's discussion in the *Laboratory Manual* of how to proceed with experimental investigations of the affective dimension of human mental life. As we begin, the open experimental question is put forward, namely "whether or not there are any other kinds of elementary mental processes" (Titchener 1901–1905, p. 91). The question is not whether it even makes sense to talk about mental elements at all, but whether the present table of mental elements requires the addition of a new theoretical category or not. Titchener's own view is that affection is a conscious element, "distinct from and ranged alongside of sensation in the composition of consciousness" (Titchener 1901–1905, p. 91),

[15] As he also put it: "I am convinced, however, that the right way to approach the study of psychological method is to assume that it is, in all essentials, identical with the observational procedure of the natural sciences" (Titchener 1912b, p. 487).

and characterized by "two qualities only, those of pleasantness and unpleasantness" (Titchener 1901–1905, p. 91).

5.6 Analysis is Its Own Test

The willingness of Titchener to subject his own view to experimental investigation clearly signals to incoming students of psychology that this is a scientific system very much amenable to experimental confirmation or disconfirmation, even on a fundamental issue such as the potential inclusion of a whole new category of elements. However, students making this inference would have failed to recognize that by the very formulation of the scientific problem, the crucial notion of mental elements has been immunized from criticism. After all, any experimental outcome, whether it is "yes, affection is a conscious element alongside sensation" or "no, affection is an attribute of sensation, not a conscious element alongside sensation," serves to reaffirm elementism. Disconfirmation of elementism is a card that was removed from the deck before the first hand was dealt.[16]

We see this when Titchener discusses the relevant "opposed opinions" to his own view of affection as a basic element. These opposing opinions (in the order he mentions them) are the theories according to which (1) "affection is merely an attribute of sensation," (2) "affection is simply a closely welded complex" of sensations, or (3) affection is a special type of sensation (Titchener 1901–1905, p. 91). The fundamentally opposed opinion, that the mental is not at all to be understood in terms of elemental empiricist sensations at all, is the epistemological elephant in the room. The notion of introspection at work, here, is a highly idiosyncratic and theory-laden one, with elementism residing at the very core.

To illustrate another instance of how this works, consider a randomly chosen experiment (experiment 21) from the *Laboratory Manual*, where the aim is to "determine the relative affective value of colored impressions" (Titchener 1901–1905, p. 92). Here, the subject is shown a series of 27 colored pieces of paper in 351 separate two-color combinations and is asked to judge which is the more and which is the less pleasant of the two. The experiment "consists in the comparison of the affective value (pleasantness or unpleasantness) of every color with that of every other color in the series" (Titchener 1901–1905, p. 93). The results are tabulated into the form of a curve and the experimenter then attempts to gather a uniformity of preference from this.

The methodological problems with the experimental design include the incorporation of Titchener's own favored view of affection ("two qualities only," pleasantness

[16] This is a state of affairs radically different from normal experimentation. Normally, the experimental setup and the interpretation of the experimental results *are* based on a number of theoretical presuppositions. However, and this is the crucial difference, these theoretical presuppositions are not themselves completely immunized from scientific *disconfirmation*.

and unpleasantness) into the very response options of the subject. But much worse still, any experimental results of this type cannot, by the nature of the setup, do anything but strengthen the assumptions of elementism, reductionism, and sensationism. At most, the "opposed opinions" would be confirmed in which case elementism still is assumed and affection simply is reduced to either a quality, a compound, or a type of sensation.

Now consider how we *validate* the results we obtain from our experiments. In *Outline of Psychology*, Titchener discusses two ways in which such a dataset "needs to be tested" (Titchener 1896, p. 14). First, we must always ask whether our analysis has gone as far as it can go. Has it taken account of all the elements that are contained in the experience? To answer this question, "the analysis must be repeated: *analysis is its own test*" (Titchener 1896, p. 14, italics added). If one psychologist claims that a process is elemental, other psychologists should perform the analysis as well to see whether they can go further or whether it is indeed elemental. In short,

> we analyze again and again; and if the result is always the same, we are satisfied to let it stand. Children who do not know how to prove an example in arithmetic follow the same plan; if they get the same answers several times over, and if their schoolmate gets that answer too, they are satisfied; and when the work has been honestly done, the agreement is pretty good evidence that they are right (Titchener 1920a, pp. 16–17).

Second, the analysis can also be tested by synthesis, by putting the components back together to see whether we can reconstruct the whole experience.

> When we have analyzed a complex into the elements a, b, c, we test our analysis by trying to put it together again, to get it back from a, b, and c. If the complex can be thus restored, the analysis is correct; but if the combinations of a, b, and c does not give us back the original complex, the analyst has failed to discover some one or more of its ingredients" (Titchener 1896, p. 14).

The defectiveness of this should be apparent. Peer confirmation (or disconfirmation) of data is indeed an important feature of a proper experimental methodology but the procedure that Titchener offers under this guise is counterfeit. Presumably all the peer psychologists attempting the analysis and synthesis replication began their studies of experimental psychology with Titchener's *Manual* in hand; they all belonged to the same overall school of thought in psychology, and they all accepted the method of psychological analysis as fundamental to psychological study. If the analysis and synthesis of the mental elements is confirmed or disconfirmed by criteria such as whether other psychologists "can go further" in the process, the question of the validity of the basic approach is never raised.

Titchener's positive and negative strategy insulated his fundamental empiricist assumptions from experimental disconfirmation as well as from scientific criticism. The issue can be illustrated by Titchener's example of children arriving at an answer in arithmetic: if the children are recognized as authorities on mathematics, as Titchener and his followers were in their respective field at the time, and if they arrive at the same answers because they all employ the method of locating the same misprinted answers in the back of the arithmetic textbook, then we have a fruitful soil out of which invalid experimental findings, like weeds, can proliferate freely.

References

Berman D, Lyons W (2007) The first modern battle for consciousness: J. B. Watson's rejection of mental images. J Conscious Stud, 14(11):4–26
Boring EG (1921) The stimulus-error. Am J Experimental Psychol, 32(4):449–471
Boring EG (1927) Edward Bradford Titchener: 1867–1927. Am J Psychol 38(4):489–506
Boring EG (1937) Titchener and the existential. Am J Psychol 50(1/4):470–483
Costall A (2006) 'Introspectionism' and the mythical origins of scientific psychology. Conscious Cogn 15(4):634–654
Danziger K (1980) The history of introspection reconsidered. J Hist Behav Sci 16(3):241–262
Danziger K (2001) Introspection: history of the concept. In: Smelser NJ, Baltes PB (eds) International encyclopedia of the social and behavioral sciences, Elsavier Science, Pergamon, pp 7888–7891
English HB (1921) In aid of introspection. Am J Psychol 32(3):404–414
Evans RB (1990) The scientific and psychological positions of E. B. Titchener. In Evans and Leys (eds) Defining American psychology: The correspondence between Adolf Meyer and Edward Bradford Titchener. Johns Hopkins University Press, Baltimore
Goldman AI (2004) Epistemology and the evidential status of introspective reports. J Conscious Stud 11(7–8):1–16
Schwitzgebel E (2002) How well do we know our own conscious experience? The case of visual imagery. J Conscious Stud 9(5–6):35–53
Schwitzgebel E (2004) Introspective training apprehensively defended: reflections on Titchener's Lab Manual. J Conscious Stud 11(7–8):58–76
Schwitzgebel E (2005) Difference tone training: a demonstration adapted from Titchener's Experimental Psychology. Psyche 11(6)
Schwitzgebel E (2007) Eric Schwitzgebel's contribution to Describing inner experience: Proponent meets skeptic (Hurlburt and Schwitzgebel), MIT Press, Cambridge
Schwitzgebel E (2011) Perplexities of consciousness. MIT Press, Cambridge
Titchener EB (1896) Outline of psychology. The Macmillan Company, New York
Titchener EB (1899) Structural and functional psychology. Philos Rev 8(3):290–299
Titchener EB (1901–1905) Experimental psychology: a manual of laboratory practice. The Macmillan Company, New York
Titchener EB (1910) The past decade in experimental psychology. Am J Psychol 21(3):404–421
Titchener EB (1912a) Prolegomena to a study of introspection. Am J Psychol 23(3):427–448
Titchener EB (1912b) The schema of introspection. Am J Psychol 23(4):485–508
Titchener EB (1914) A primer of psychology. The Macmillan Company, New York
Titchener EB (1920a) A beginner's psychology. The Macmillan Company, New York
Titchener EB (1926) A text-book of psychology. The Macmillan Company, New York

Chapter 6
The Imageless Thought Controversy

6.1 Bewußtseinslagen

Titchener's "introspective" methodology, then, was afflicted with profound congenital defects, rendering it scientifically unviable. For Titchenerian psychology, the highly visible *symptom* of underlying pathology came in the form of the imageless thought controversy. Martin Kusch describes the imageless thought controversy as follows:

> [It] was triggered by methodological and theoretical innovations coming out of the Psychological Institute of the University of Würzburg between, rough, 1900 and 1907; the debate … peaked between 1907 and 1912.
>
> The prominence given to this debate in almost all histories of psychology attests to the fact that the Imageless Thought Controversy was perhaps the decisive juncture in the history of psychology and, thus one of the key events in twentieth-century science. (Kusch 1995, p. 420).

Oswald Külpe (1862–1915), director of the Institute at Würzburg from 1896 to 1909, was the recognized leader of the Würzburg side of the controversy. As far as the genesis of the Würzburg investigative program, according to Boring "[t]he Würzburg school begins technically" with the 1901 paper by Mayer and Orth titled "Zur qualitativen Untersuchung der Assoziation," published in *Zeitschrift für Psychologie und Physiologie der Sinnesorgane* (1950, p. 402). This research seemed to show that the existing scheme of mental classification was incomplete.[1] In Mayer and Orth's words, in addition to the psychological categories of images and volitions,

> we must set up also a third group of facts of consciousness, one that has not been sufficiently stressed in psychology up to the present day. In the course of our experiments we were, again and again, involuntarily brought up against the fact of the existence of this third group. The subjects frequently reported that they experienced certain events of

[1] Mayer and Orth 1901, pp. 1–13.

C. Beenfeldt, *The Philosophical Background and Scientific Legacy of E. B. Titchener's Psychology*, SpringerBriefs in Philosophy, DOI: 10.1007/978-3-319-00242-2_6,

> consciousness which they could quite clearly designate neither as definite images nor yet as volitions (Mayer and Orth 1901, p. 6).[2]

The controversy arose because the Würzburg psychologists claimed to have discovered a new type of basic and irreducible mental constituent, variously referred to as "*Bewußtseinslagen*" (conscious sets), "*Bewußtheiten*" (awarenesses) or, simply, "*Gedanken*" (thoughts).[3] These purported mental phenomena appeared in research on judgment, studies of the will, and investigations of the psychology of thought.[4] In the Würzburg laboratory, a form of introspective analysis repeatedly showed (or, at least, appeared to show) that "knowledge exists in an imageless form, that is, no phenomenological components are demonstrable—neither visual, acoustic, nor kinesthetic sensations, nor their memory images—which would qualitatively define the content of this knowledge" (Ach 1905, p. 25).[5] "Psychology," as Boring explained, "was ready to find that thinking is not all sensory" (Boring 1950, p. 406). However, this position was utterly contrary to Titchener's convictions. He believed in a robustly sensationist account of the nature of the elements of mind.[6] According to this view, sensations are the basic constituents of perceptions,[7] of mental images (which differ from sensations in degree, not kind), and of ideas.[8] "Sensation," Titchener declared, "is the raw material from which ideas are built up" (Titchener 1895, p. 428).

6.2 Dogmatic Affirmation and Denial

The problem was that, to many psychologists, the Würzburg investigations of imageless thoughts seemed to arise from a *correct employment* of Titchenerian methodology, i.e. of psychological analysis, by seasoned experimenters. In the paper "Imageless Thought," published by *The Psychological Review* in 1911,

---

[2] The translated passage is from Humphrey (1951), p. 33.

[3] We here adopt Kusch's translation of these terms (Kusch 1995, pp. 421–422). Humphrey translates "*Bewußtseinslagen*" as "states of consciousness" (Humphrey 1951, p. 33).

[4] Chronologically: Karl Marbe, *Experimentell-psychologische Untersuchungen über das Urteil* (Leipzig: Engelmann, 1901); Johannes Orth, Gefühl und Bewusstseinslage (Berlin: Reuther and Reichard, 1903); August Messer, *Experimentell-psychologische Untersuchungen über das Denken*, Arch. Gesamte Psychol., 1906, 8:1–224; Narciß Ach, *Über die Willenstätigkeit und das Denken* (Göttingen: Vandenhoeck and Ruprecht 1905).

[5] The translated passage is from Rapaport (1951), p. 24.

[6] E.g. Titchener (1909).

[7] "A perception … is primarily a group of sensations arranged by external nature," as he put it in one formulation (Titchener 1901–1905, p. 129).

[8] An idea, as Titchener maintained in the *Text-Book*, "differs from a perception only by the fact that it is made up wholly of images" (Titchener 1926, p. 376).

Angell (an opponent of the notion of imageless thought) explains the state of play at the time. Briefly put, the conflict had ground to a halt and the condition was that of stalemate. After examining the relevant literature, as Angell describes it,

> one is left with the feeling that the case is largely reduced to mere assertion and denial, occasionally to vituperative recrimination. It seems to be largely a matter of "It is!" or "It isn't", adorned with such adjectives as taste may dictate and capacity afford (Angell 1911, p. 305).

"Competent introspectionists" are thus, as Angell put it, "arrayed on the two sides of the question, and the results which they bring in are equally unequivocal and equally dogmatic" (Angell 1911, p. 305). Not just a few individuals, however, but *numerous* authorities were to be found on each side. This, in turn, meant that the imageless thought side could not simply be ignored.

So long therefore as the number of persons remains small, who have been trained in introspective methods, and who report affirmatively upon the presence of imageless thought, it will always seem a reasonable interpretation that errors of observation are responsible for the discrepancy between their reactions and those of trained observers. But the number of individuals who are now to be counted among those making affirmative allegations is suspiciously large to justify this interpretation (Angell 1911, p. 306).[9]

The controversy, then, had brought psychological analysis to a fork in the road where both directions ahead seemed highly uninviting. Going right, one would continue with the Titchenerian strategy of dismissing the imageless thought findings as arising from the stimulus error. Yet, as Angell observes, an "issue of this character is hardly likely to be settled merely by dogmatic affirmation and denial, which is the level of the present characterizing the controversy" (Angell 1911, p. 313). Going left, one would explain the irresolvable conflict by simply asserting that the "imageless thought" researchers are differently constituted at the level of basic biology, "comparable in character" to "a race of men having six fingers instead of five on each hand, or having two noses or three eyes" (Angell 1911, p. 311).

In other words, if we reject Titchener's stimulus error strategy and also reject the idea of dividing the human race into two classes (depending on their relative proximity to either the Cornell or the Würzburg laboratory), it seemed we must

> simply wait for the amassing of evidence in the hope that the accumulation of experimental results will slowly create a presumptive proof for the position of one or other of the contestants (Angell 1911, p. 313).

Because psychological analysis was built on a foundation of mistaken assumptions—namely, elementism, decompositional reductionism, sensationism, and the

[9] "[A]lthough, as Angell adds by way of rather unconvincing pragmatism, "the author is at present obliged, as the lesser of two evils, to believe that this line of explanation is essentially correct" (Angell 1911, pp. 306–307).

laws of association—and because it immunized those assumptions from experimental correction, this means that Titchener's idiosyncratic methodology had, as we might put it, finally pushed the science of psychology into the Theatre of the Absurd.

6.3 Waiting for Godot

In one of the most prominent plays in the absurdist tradition, Samuel Beckett's post-World War II play *Waiting for Godot*, we see Vladimir and Estragon wait interminably for the appearance of Godot. To pass the time, they converse, theorize, argue, play games, and contemplate suicide. Godot never appears. In a similar manner, the *decisive* experimental result that Angell in 1911 suggests the psychology profession might wait for would not and could not appear either—not, that is, within the Titchenerian framework of psychology.[10] Unlike Vladimir and Estragon, however, the psychological profession was not equipped with interminable patience.[11] On June 28th 1914, shots fired in Sarajevo ignited a worldwide military conflagration. There was little patience for leisurely psychological fiddle-playing while Europe burned. The tired stalemate was broken by a brash young psychologist studying animal behavior at Johns Hopkins University. Quoting again his famous 1914 declaration,

> two hundred years from now, unless the introspection method is discarded, psychology will still be divided on the question as to whether auditory sensations have the quality of "extension," whether intensity is an attribute which can be applied to color, whether there is a difference in "texture" between image and sensation; and upon many hundreds of others of like character (Watson 1914, p. 8).

The ad hoc nature of psychological analysis brought this state of affairs about—but the consequence was that *introspection as such* was blamed and, ultimately, discredited in the minds of a whole generation of young experimental psychologists. We see this move towards a general dismissal of introspection when we consider

> Titchener's criticism of the findings of imageless thought. The findings, [Titchener] says, are not convincing, because the observers were all under the influence of the 'stimulus error' in their introspective reports, that is, they have reported *about* their

[10] As Holt observed, the "critique usually made, that the introspective method gave different results misses the main point; even if there had been no inconsistency, the end of the line had been reached at a point" where Titchener's method of "introspection yielded nothing that made any sense" (Holt 1964, p. 256).

[11] Titchener seems to have thought, to the very end, that he could furnish an analytic decomposition that would settle the matter. As he kept on insisting, "unanalyzed is not unanalyzable" (Evans 1975, p. 341; Evans is here drawing upon personal communication with Karl Dallenbach in 1963).

experiences, without really describing the contents they experienced (Ogden 1911, pp. 192–193).

But if the error be regarded so serious as to discredit the entire result of Bühler's investigation, as Titchener appears to believe, may we not carry the point a step farther and deny the value of all introspection. Indeed, in a recent discussion among psychologists, this position was vigorously maintained by two among those present (Ogden 1911, p. 193).

There was nothing inherently significant about this particular controversy. It *became* significant because it was a symptom that was highly visible to the psychological profession in general—and to young ambitious psychologists like John B. Watson in particular—of profound methodological pathology in the field.

6.4 Other Symptoms

The imageless thought controversy was only one visible symptom; it was far from the only serious defect of Titchenerian psychology.[12] To examine some of these flaws, let us start again from the Titchenerian perspective on psychology. "The conscious process" Titchener tells us, is like "a fresco, painted in great sweeps of color," and "words are little blocks of stone, to be used in the comparison of a mosaic." If we are "required to represent the fresco by a mosaic, we must see to it that our blocks be of small size and of every obtainable tint and hue. Otherwise, our representation will not come very near to the original" (Titchener 1896, p. 36). According this approach, then, one recognizes some 40,000 elemental items, any number of which at a given moment in time may "flow together, mix together, overlapping, reinforcing, modifying or arresting one another"[13] (Titchener 1896, p. 15) to create the fantastic complexity of experience—and one sets out to use fresco block words to capture isolated sensations, never things.

The result, in short order, is a research program that soon inflates to unbounded proportions. Undertaking the task of unraveling one day in the life of Ivan Denisovich, as it were, into all of its moment-to-moment sensory constituents and the myriad causal interactions of these constituents almost seems to dwarf the Space Program in terms of sheer scope. Yet, even if a list of the

[12] Koffka identified the larger family of problems, of which the imageless thought controversy was just one member: "What are we to do if the introspections of different observers do not agree? There are many examples of this in psychology. The disputes about the imageless thought, about the so-called attributes of sensation, about the primary colors, are all cases in point. The traditional theory of psychological method cannot adduce a criterion to decide which description is the true one" (Koffka 1924/1925, p. 152).

[13] In obedience to various psychological laws of association and so on.

elements and their various interactions were somehow to be delivered, we would still find that, just like Humpty Dumpty after his fall, all the king's horses and all the king's men would be unable to put the atomic sensations back together again—together, that is, in the form of genuine human perception, cognition, thought, memory and emotion.

Consider, for example, the attempt to bring back even a single day of normal life—of reading the newspaper, hurrying to work, drinking coffee, having meetings, doing research, solving problems, making budgets, writing e-mails, remembering your father's birthday, driving across town during rush hour, buying groceries, picking up the kids, having dinner, planning your vacation, watching the sunset, meeting friends, discussing politics, going home, watching a film, falling asleep—from a countless mass of splintered elemental processes, without getting into the meaning, purpose, logic and value significance of anything. This splintering represents, in Danziger's apt words, the "Titchenerian destruction of actual lived experience" (Danziger 1990, p. 47).

Even on its own terms, then, introspectionism failed. As William James said about the work of such prism, pendulum, and chronograph-philosophers, "in some fields the results have as yet borne little theoretical fruit commensurate with the great labor expended in their acquisition" (James 1952, pp. 127).[14] More bluntly, Boring observed that introspection "with inference and meaning left out as much as possible," i.e. introspection understood as psychological analysis, "becomes a dull taxonomic account of sensory events which, since they suggest almost no functional value for the organism, are peculiarly uninteresting to the American scientific temper" (Boring 1953, p. 174).

Titchener advanced a science dedicated to the discovery, identification, and systematic organization of mental elements and molecules, a "pure" science, removed from the necessity of practical application.[15] As he saw it, this was the beginning, the middle, and the end of genuinely scientific psychology. Like a chemist occupied by the task of mapping out the basic chemical elements in a carefully regulated laboratory setting, there was no need to go into the field and study the messy compounds of everyday life.

[14] He adds, though, that "facts are facts, and if we only get enough of them they are sure to combine" (James 1952, p. 127).

[15] "[M]y own standpoint," as he put it, "is that of pure science, or the desire for knowledge without regard for utility" (Titchener 1910, p. 405). And, again, in *Science* in 1919, he states in response to Thorndike, "I believe very strenuously in pure science" (Titchener 1919, p. 170).

By deliberate intent, Titchener's psychology lived in splendid isolation from animal psychology, developmental psychology, and psychopathology.[16, 17] Human psychology, Titchener tells us, "confines itself to the human consciousness" (Titchener 1896, p. 17), by which he meant the normal, mature human mind (Titchener 1905, p. 208). What we today might call neuropsychology he regarded as outside the concern of the psychologist. He was "not willing to discuss the nervous system, on the assumption that it was properly the topic of the physiologist" (Pillsbury 1929, pp. 274–275).

While there were sound scientific reasons to reject the dynamic Freudian unconscious, Titchener rejected any talk of mental processes that are not conscious.[18] Conscious observation of an inherently unconscious or subconscious process is of course impossible, but Titchener maintained "that there is no mental process that cannot be observed experimentally" (Titchener 1914, p. 32). Observed, in this case, meant consciously experienced by a subject in an appropriate experimental setting, and without such observation it follows that there is "no psychological evidence of a mind which lies behind mental processes" (Titchener 1896, p. 341).

The insights from Darwinian and comparative biology were also, on the whole, theoretically alien to introspectionism since Titchener's approach, via associationism,

[16] Boring tells us, for example, that Titchener "rejected animals, children and clinical patients as psychological subjects, because they could not introspect" (Boring 1950, p. 419). We should recognize, however, that Titchener's own stated position on the matter was more complex. First, he was certainly aware that animal, developmental, and abnormal psychology were becoming increasingly popular fields of investigation. Second, he encouraged a certain kind of cooperation between investigators of these fields and proper laboratory psychologists. "We can hardly open a magazine nowadays," as he put it, "without finding applications of the experimental beyond the limits of the normal, adult, human mind. In animal psychology, in child psychology, in various departments of mental pathology, the experimental method is employed" (Titchener 1905, p. 221). Considering here just animal psychology, he urges "more cooperation between the … student of comparative psychology, and the laboratory psychologist" (Titchener 1905, p. 223). Yet, he also voices significant concerns about these novel areas of psychological investigation. These include the "crudity and roughness of the work," the risk of "overhasty generalization," the inferential (rather than directly introspective) nature of comparative psychology, and the fact that even for this kind of work "introspection gives the pattern, sets the standard of analysis and explanation." Further we "can hardly hope," as he foresees it, that the novel and the traditional psychological interests "will be combined in the same person." As he puts it, "[w]hen one has once stepped inside the ring of the normal, adult consciousness, there is very little temptation to step out again; the problems …. are enough to occupy several generations of workers" (all quotes: Titchener 1905, p. 223) Titchener offers, then, not a simple rejection of these fields, but a clear downplaying of them as serious competitors to proper laboratory psychology.

[17] For similar reasons, it seems, Titchener was also opposed to psychological testing (see Freeman 1984).

[18] In an early paper he even makes this a matter of definition: "consciousness is the collective name of the mental processes which exist for an individual at any given moment" (Titchener 1893, p. 452). Unconscious mental processes are thus, by this definition, not possible. In British associationist thought, John Stuart Mill similarly held unconscious mental processes as impossible (Hatfield 2003, p. 95).

was built according to the thoroughly non-organic blueprint of Newtonian physics and nineteenth century inorganic chemistry (Titchener 1925, p. 313). It was an ivory tower celestial mechanics of the mental, essentially untouched by the incoming data about human and animal biology.[19] Titchenerian psychology thus failed at the task it had designated for itself, namely, to scientifically study the nature of the mental. Its tenacious adherence to the premises of elementism, reductionism, and sensationism along with the fundamental methodological assumption of analysis, finally meant that all substantive ties to the other sciences were severed and that the psychological issues of interest to ordinary people were ignored.

6.5 The Final Demise

Introspectionism, of course, continued as an influential research program well into the first two decades of the twentieth century but then plummeted;[20] by the 1930s it had fallen into general disrepute and was completely abandoned.[21] The reason was that its shortcomings, at this point, had become evident for all to see. It was "applicable only to laboratory situation," as Ulric Neisser observed, and it "lacked any clear account of how people interact with the world." Further, it had "no theory of cognitive development," it had "no theory of unconscious processes," it offered "no serious theory of behavior" (Neisser 1976, pp. 2–3)—and, in short—"introspective psychology left out nearly everything that ordinary people thought important." It is "not surprising, then, that it was abandoned in favor of more promising ideas" (Neisser 1976, p. 3).

If one observed this state of affairs, a conclusion would seem to follow readily, namely that introspection should never again find a place in scientific psychology. This idea became a foundational assumption and guiding principle for behaviorism. As Watson famously declares in 1914, on the first page of *Behavior: An Introduction to Comparative Psychology*, psychology as the behaviorist views it

> is a purely objective experimental branch of natural science. Its theoretical goal is the prediction and control of behavior. Introspection forms no essential part of its methods, nor is the scientific value of its data dependent upon the readiness with which they lend themselves to interpretation in terms of consciousness (Watson 1914, p. 1).

As he explained, one must

> believe that two hundred years from now, unless the introspection method is discarded, psychology will still be divided on the question as to whether auditory sensations have the quality of "extension," whether intensity is an attribute which can be applied to color, whether there is a difference in "texture" between image and sensation; and upon many hundreds of others of like character (Watson 1914, p. 8).

---

[19] Although Titchener sometimes spoke of the structural analysis of mind as analogous to the structural analysis of organisms (e.g. Titchener 1898, pp. 449–451).

[20] In one estimate, the decade from 1903 to 1913 can be pinpointed as the period of greatest flourishing and proliferation of classical introspectionism (Danziger 1980, p. 255).

[21] English (1921) is a good example of a late-stage defense of Titchenerian methodology.

Titchener's method was, quite mistakenly, taken to simply be "the introspection method," (as Watson termed it in the quote above), and so the inference follows readily that psychology will make little progress if it continues to rely on introspection. Introspection, consequently, should be banned from scientific psychology.

> Historically … it was thought that introspection was shown to generate massive contradictions, thereby revealing its substantial unreliability (Goldman 2000, p. 19).

> Historically, psychology challenged the legitimacy of introspection as a method of psychological science. This challenge was so potent during the behaviorist era that the very word introspection became—and largely remains—taboo. … (Goldman 2006, p. 228).

The historical assessment that Watson made regarding the failure of introspectionism is largely correct. Introspectionism (and, for much the same reason, associationism) were scientific failures. The error lies in the first premise, namely that introspectionism employed a form of enquiry even remotely close to introspection *simpliciter*. The terms "experimental introspection," "scientific introspection," and sometimes just plain "introspection" were *actually* used to denote a Byzantine and heavily theory-laden procedure of "psychological analysis" or regimented "analytic attention" designed to generate data confirming a set of assumptions regarding mental ontology taken over from British associationism, while at the same time shielding those assumptions from disconfirming evidence.

References

Ach N (1905) Über die Willenstätigkeit und das Denken. Vandenhoeck & Ruprecht, Göttingen

Angell JR (1911) Imageless thought. Psychol Rev 18(5):295-323

Boring EG (1950) A history of experimental psychology. Appleton-Century-Crofts, New York

Boring EG (1953) A history of introspection. Psychol Bull 50(3):169–189

Burnham JC (1968) On the origins of behaviorism. J Hist Behav Sci 4(3):143–151

Danziger K (1980) The history of introspection reconsidered. J Hist Behav Sci 16(3):241–262

Danziger K (1990) Constructing the subject: historical origins of psychological research. Cambridge University Press, Cambridge

English HB (1921) In aid of introspection. Am J Psychol 32(3):404–414

Evans RB (1975) The origins of Titchener's doctrine of meaning. J Hist Behav Sci 11(4):334-341

Freeman FS (1984) A note on EB Titchener and GM Whipple. J Hist Behav Sci 20(2):177–179

Goldman AI (2000) Can science know when you're conscious? Epistemological foundations of consciousness research. J Conscious Stud 7(5):3–22

Goldman AI (2006) Simulating minds: the philosophy, psychology, and neuroscience of mindreading. Oxford University Press, Oxford

Hatfield G (2003) Psychology: old and new. In: Baldwin (ed) The Cambridge history of philosophy 1870–1945. Cambridge University Press, Cambridge

Holt RR (1964) Imagery: the return of the ostracized. Am Psychol 19(4):254–264

Humphrey G (1951) Thinking: an introduction to its experimental psychology. Methuen, London

James W (1952) The principles of psychology. Encyclopædia Britannica, Chicago

Koffka K (1924/1925) Introspection and the method of psychology. Br J Psychol 15(2):149–161

Kusch M (1995) Turn-of-the-century psychological research schools. Isis 86(3):419–439

Mayer A, Orth J (1901) Zur qualitativen Untersuchung der Association. Zeitschrift für Psychologie und Physiologie der Sinnesorgane 26:1–13

Neisser U (1976) Cognition and reality: principles and implications of cognitive psychology. W. H. Freeman & Company, San Francisco
Ogden RM (1911) Imageless thought: résumé and critique. Psychol Bull 8(6):183–197
Pillsbury WB (1929) The history of psychology. W. W. Norton & Company, New York
Rapaport D (1951) Organization and pathology of thought. Columbia University Press, New York
Titchener EB (1893) Two recent criticisms of `modern' psychology. Philos Rev 2(4):450–458
Titchener EB (1895) Psychology. Science, New Series 1(16):426–431
Titchener EB (1896) Outline of psychology. The Macmillan Company, New York
Titchener EB (1898) The postulates of a structural psychology. Philos Rev 7(5):449–465
Titchener EB (1901–1905) Experimental psychology: a manual of laboratory practice. The Macmillan Company, New York
Titchener EB (1905) The problems of experimental psychology. Am J Psychol 16(2):208–224
Titchener EB (1909) Lecture on the experimental psychology of the thought-processes. The Macmillan Company, New York
Titchener EB (1914) A primer of psychology. The Macmillan Company, New York
Titchener EB (1919) Applied psychology. Science 49(1259):169–170
Titchener EB (1925) Experimental psychology: a retrospect. Am J Psychol 36(3):313–323
Titchener EB (1926) A text-book of psychology. The Macmillan Company, New York
Watson JB (1914) Behavior: an introduction to comparative psychology. Henry Holt and Company, New York

Chapter 7
Psychological Analysis: *Not* Introspection *Simpliciter*

7.1 Analogy: "Hydro-Monism"

If Titchener, as we argued above, rejected so many sources of important insight and areas of crucial practical application of psychology, did he at least have the benefit of psychological data derived from introspection? The present claim is that he did not. Despite the many calls for experimental introspection, Titchener had little regard for what we today would recognize as introspection. It is not the case, of course, that he never talked and wrote about introspection (he most certainly did), but what he actually *meant* by introspection as an investigative methodology was *in fact* a systematic experimental procedure of decompositional analysis applied, as it happens, to human psychology.

To appreciate the significance of this point, it might be helpful to consider the following analogy. Imagine that a theoretician advances *hydro-monism*, a modern version of the ancient Milesian claim that everything, *everything*, is reducible to water in its various states of matter. Imagine, further, that our imagined New Milesian system is constructed along the lines that we have here seen define Titchener's psychology. The basic assumptions include the claims that (1) everything can be reduced to water, that (2) to be scientific within this domain of investigation means nothing other than to experimentally decompose things to water, and that (3) if some experiment seems to reveal that a compound cannot be resolved into water, this shows that there must have been a basic error of methodology (the stimulus error) that should be remedied with training in *hydro-spection*, a special kind of visual perception by means of which one finally realizes that all things consist of water.

Predictably, hydro-monism generates poor scientific results and is subsequently abandoned. Premises (1), (2) and (3) explain the scientific failure of hydro-monism—*but what about visual perception*? Suppose one holds that the failure of New Milesianism demonstrates the inherent fruitlessness of visual perception because, after all, hydro-spection formed an integral part of the failed theory, and this methodology did involve (to make the example humorously absurd) staring

C. Beenfeldt, *The Philosophical Background and Scientific Legacy of E. B. Titchener's Psychology*, SpringerBriefs in Philosophy, DOI: 10.1007/978-3-319-00242-2_7,

for hours at an object until one's eyes start to water and the aqueous nature of reality becomes evident. To discount visual perception as a valid form of awareness on these grounds, however, or even to declare it unsuitable for scientific investigation, does not follow.

The present claim is that introspectionism is similar to hydro-monism and that Titchnerian introspection should no more be taken to disprove the basic human capacity for introspective self-awareness than hydro-monism should be taken to disprove our biological capacity to discriminate features of our environment by means of lens-focused reflected light. Staring for hours on end without blinking and then interpreting profuse lacrimal gland secretions as evidence for a New Milesian metaphysics is an aberrant methodological approach, not an instance of visual perception being properly exercised. Titchenerian analysis was a similarly aberrant practice in psychology.

7.2 A *Speculative* Science

The prime textbook of associationism and the work that put Titchener "on the introspective track" (Titchener 1920a, p. vii) was James Mill's *Analysis*—a work that itself was the fruit of little observation and much in the way of armchair theorizing. "Although in some ways a *tour de force*," as Hearnshaw put it, "Mill's *Analysis* is essentially a speculative construction supported by analytic reasoning and casual introspective observation" (Hearnshaw 1964, p. 2). Despite this, John Stuart Mill—following his father's approach to psychology, "had taken on the role of chief defender of the central status which introspection had always been accorded in British psychology" (Danziger 1980, p. 243). John Stuart Mill died in 1873 and two decades later Titchener was ready to take the baton.

Like associationism, Titchenerian experimental psychology was, at heart, a *speculative* rather than an observational endeavor. Countless "observations" of sorts were made, of course, but these were furnished by experimentalists trained to find only what was, on a priori, theoretical grounds, already assumed to exist—namely a mosaic-like mental world of sensory atoms. Titchener was, as we have seen, an experimental associationist at heart. The associationist theory of introspection *required* introspection to reveal the mental atoms and their lawful patterns of interaction. If introspection did *not* reveal atoms, on this view, psychology could never succeed in its aspirations to become a good Newtonian science.

Titchener substituted for anything plainly resembling what we today would call introspection the exercise of a highly idiosyncratic brand of experimental methodology, in which suitably trained experimentalists-cum-subjects search for mental elements whose existence they were instructed by the *Laboratory Manual* to assume beforehand. Auxiliary control mechanisms like the stimulus error strategy ensured that only acceptable data were scientifically admissible and that competing attempts to use introspection in psychology could be dismissed as "scientifically illegitimate."

The classical introspectionist approach was thus "bound up with analytic attention" (Koffka 1924/1925, p. 151). As Titchener explained in "The Schema of Introspection," (1) "introspection always presupposes the point of view of descriptive psychology" (Titchener 1912b, p. 508) and (2) "[d]escriptive psychology must begin with analysis, because analysis is the first task that a given subject-matter assigns to science" (Titchener 1912b, p. 500). Rather than providing a report of mental phenomena taken at their face value, "introspection in the proper laboratories always yielded sensory elements because that was 'good' observation" (Boring 1953, p. 172).

The trained experimental subjects, then, were not like photographic film or like microscopic lenses. Rather, they were more like *inference machines*, experimental automatons delivering analytic "observations," i.e. reports that were, invariably, mere confirmations of the associationist doctrine according to which they were programmed. Boring does not follow us quite this far in the evaluation of Titchenerian psychology, but he does grant that it was

> [n]ever wholly true that introspection was photographic and not elaborated by inference or meanings. Reference to typical introspective researches from Titchener's laboratory establishes this point... There was too much dependence upon retrospection. It could take twenty minutes to describe the conscious content of a second and a half and at the end of that period the observer was cudgeling his brain to recall what had actually happened more than a thousand seconds ago, relying, of course, on inference (Boring 1953, p. 174).

If there are no indivisible mental sensational elements to be found in normal human mental life, the whole Titchenerian project is misbegotten. Titchener held that scientific advance is "impossible apart from theoretical preconceptions" (Titchener 1912a, p. 438). This is true. However, it is also true that if those preconceptions turn out to be entirely wrong, scientific advance will not follow.

7.3 Repudiating the Facts of Introspection

Criticism of this approach to psychology was forcefully advanced by the Gestalt psychologists. In Koffka's words, the classical introspectionist approach treats mental content as if it were "a material thing," yet,

> whatever the mind is, it is not a mosaic of solid unalterable things. Here we are face to face with what is to my mind one of the basic errors of traditional psychology. On this error the whole system rests. This error is the justification of the selection we are to make among our contents, and it leads consistently to the fundamental concepts of sensation, image and feeling as *the* mental elements. Reality, as it appears to each of us, is to be reduced to these simple terms, although the richness of our world transcends these limits and the purposiveness of our own mind may oppose our efforts after simplification (Koffka 1924/1925, p. 151).

Koffka argues, correctly, that a "science built upon introspection ought to acknowledge these difficulties," and, "*by not doing so it vitiates its own principles, over and over again repudiating facts of introspection and explaining them away*" (Koffka 1924/1925, p. 151; italics added).

Indeed, Titchener warns against attempting to furnish a "phenomenological account of mind" by which he means "an account which purports to take mental phenomena at their face value" by recording "them as they are 'given' in everyday experience" as furnished by a "naïve, common-sense, non-scientific observer" (Titchener 1912b, p. 489). Paradoxically, then, Titchenerian introspection delivered precisely what *no* ordinary person in ordinary circumstances *ever* comes introspectively into contact with as part of normal mental life—namely a mosaic of psychological atoms.

According to Watson, "Titchener ... has fought the most valiant fight in this country for a psychology based upon introspection" (Watson 1914, p. 8). In a similar vein, contemporary thinkers assume, as we saw in the introduction, that Titchener practiced the archetypal method of introspective psychology. Both assessments are part of the received wisdom today. In conjunction with the observation that Titchenerian psychology clearly *failed* as a scientific investigative project, we are led to the conclusion that scientific psychology can make little progress unless it excludes or ignores introspection.

This state of misapprehension and confusion is a major cognitive challenge to the methodologically confident use of introspection in scientific psychology today (see Jack and Roepstorff 2002, pp. 133–134). Put simply, it turns out that one of the most widely taught lessons in the history of psychology reflects confusion rather than clear understanding. We should instead recognize that the standard story is wrong and that Titchenerian psychology did not implode because it was guilty of an over-reliance on introspection.

Introspectionism *did* implode, of course, and it did so because there were profound structural flaws in its experimental approach, but those flaws were not caused by introspection. They were caused by the speculative associationist assumptions upon which the system rested and, in turn, by the scientifically flawed methodology that was devised accordingly,[1] a method that confusingly and regrettably was called experimental or scientific introspection by Titchener.

7.4 "Introspection" in Newspeak

Confusingly and regrettably, that is, but not purposelessly. In other words, to the extent that Titchenerian psychology relied on Orwellian Newspeak in its use of the term "introspection," this was part of a *deliberate epistemic strategy* of reading a new meaning (psychological analysis) into an old term ("introspection").

> Introspection, then, is in a peculiar and exclusive sense the business of the psychologist, and it is well that this business should have a specific name. When, moreover, we have a

[1] Watson summed this up in his 1913 paper. "Psychology, as it is generally thought of, has something esoteric in its methods. If you fail to reproduce my findings ... it is due to the fact that your introspection is untrained. *The attack is made upon the observer and not upon the experimental setting*" (Watson 1913, p. 163; italics added).

> traditional term, that is full of misleading suggestions to the student, *it is wiser to adopt that term, reading the suggestions out and reading a sound definition in*, than to pass by and introduce new coinage (Titchener 1912b, p. 488; italics added).

In other words, the

> term Introspection, as we find it used to-day, is highly equivocal ... *I reserve the name henceforth for the methods that are scientifically available and that appear to have been actually employed* (Titchener 1912b, p. 485, italics added).

The "methods that are scientifically available," as Titchener saw it, were the systematic approaches already known in the physical sciences of breaking compounds down, first into simpler parts and then finally into atomic or elemental constituents.

In the world of *Nineteen Eighty-Four*, to enjoy "goodthink" is *by definition* to think pro-Party thoughts. In Titchenerian psychology, to engage in "introspection" is, by definition, to engage in psychological analysis. About this sort of linguistic strategy Charles Lutwidge Dodgson, a contemporary of Titchener's in Oxford, wrote the following:

> "[G]lory" doesn't mean a "nice knock-down argument," Alice objected. 'When *I* use a word,' Humpty Dumpty said in a rather scornful tone, 'it means just what I choose it to mean—neither more nor less.' 'The question is,' said Alice, 'whether you *can* make words mean so many different things.' 'The question is,' said Humpty Dumpty, 'which is to be master—that's all.' (Carroll 1994, p. 100).

Schwitzgebel claims that "Titchener and other introspective psychologists ... exhorted their observers to set aside their presuppositions" and, furthermore, "generally attempted to reduce or disarm their own expectations" (Schwitzgebel 2007, p. 225).[2]

There is a sense in which this is true. For example, the methodological injunction against the "stimulus error" was, in part at least, meant to stop overt theorizing by the subjects (e.g. "Since I know the second object is more massive than the first, it must have felt heavier against my skin"). Yet, as we have seen, there is a deeper sense in which Titchenerian psychology was *all about* associationist presuppositions and expectations.

The so-called "golden age of introspection" (Lyons 1986) was misnamed. It was really an age of associationism, and, most significantly, an age of elementism, reductionism and sensationism. As far as mental philosophy was concerned, it was an age in which the scientific breakthroughs in physics and chemistry were systematically (mis)applied to the study of the mind; in this respect it was an age of mental physics. Titchenerian analysis was the jewel in the crown of this distinctive approach to psychology.

This is how we should understand introspectionism in early scientific psychology.

---

[2] In a similar vein, Gertler and Shapiro hold that Wundt *and Titchener* "championed the *judicious* use of introspection" (Gertler and Shapiro 2007, p. 58; italics added).

References

Boring EG (1953) A history of introspection. Psychol Bull 50(3):169–189
Carroll L (1994) Through the looking glass. Penguin, London
Danziger K (1980) The history of introspection reconsidered. J Hist Behav Sci 16(3):241–262
Gertler B, Shapiro L (2007) Consciousness: how should it be studied? In: Gertler B, Shapiro L (eds) Arguing about the mind. Routledge, Oxford
Hearnshaw LS (1964) A short history of British psychology 1850–1940. Methuen & Co, London
Jack AI, Roepstorff A (2002) Introspection and cognitive brain mapping: From stimulus-response to script-report. Trends Cogn Sci 6(8):333–339
Koffka K (1924/1925) Introspection and the method of psychology. Br J Psychol 15(2):149–161
Lyons W (1986) The disappearance of introspection. MIT Press, Cambridge
Schwitzgebel E (2007) Eric Schwitzgebel's contribution to Describing inner experience: Proponent-meets skeptic (Hurlburt and Schwitzgebel), MIT Press, Cambridge
Titchener EB (1912a) Prolegomena to a study of introspection. Am J Psychol 23(3):427–448
Titchener EB (1912b) The schema of introspection. Am J Psychol 23(4):485–508
Titchener EB (1920a) A beginner's psychology. The Macmillan Company, New York
Watson JB (1913) Psychology as a behaviorist views it. Psychol Rev 20:158–177
Watson JB (1914) Behavior: an introduction to comparative psychology. Henry Holt and Company, New York